Turning, Milling and Grinding Processes

Patrick Byrne BA

Section Head–Mechanical Engineering NVQ
North Devon College

ARNOLD

A member of the Hodder Headline Group
LONDON • SYDNEY • AUCKLAND

Acknowledgements

The author expresses his thanks to the following organisations:

Jones & Shipman plc, Leicester
HW Ward (Service Spares) Ltd, Cheshire

Every possible effort has been made to trace copyright holders. Any rights not acknowledged here will be acknowledged in subsequent printings if notice is given to the publishers.

First published in Great Britain in 1996 by Arnold,
a member of the Hodder Headline Group,
338 Euston Road, London NW1 3BH

British Library Cataloguing in Publication Data
A catalogue record for this book is available from the British Library

ISBN 0 340 62503 1

Produced by Gray Publishing, Tunbridge Wells, Kent
Printed and bound in Great Britain by The Bath Press, Bath

Contents

1

Health and safety

Health and safety matters are now covered by a number of laws and regulations. Some of these are very recent and are designed to harmonise health and safety practices across the European Union (EU). In most cases these new regulations replace or extend existing British industrial law.

There is a prime duty for employers to provide safe places, equipment and systems of work, and to care for the health, welfare, instruction and training of their personnel.

Employees must also take responsibility for their own health and safety and that of others, and must cooperate fully with all safety measures, procedures and instructions.

Employers must identify significant risks to health and safety arising from work activities and take steps to remove or control these risks. Employees must comply with all safety requirements, and report all dangerous situations and occurrences, including potential dangers.

Health and safety law

The following Acts of Parliament and Regulations are some of the main ones that relate to workshop practice:

1 The Health and Safety at Work Act 1974
2 The Management of Health and Safety at Work Regulations 1992
3 The Control of Substances Hazardous to Health Regulations 1988 (COSHH)
4 The Personal Protective Equipment at Work Regulations 1992
5 The Manual Handling Operations Regulations 1992
6 Workplace (Health, Safety and Welfare) Regulations 1992
7 The Provision and Use of Work Equipment Regulations 1992
8 The Factories Act 1961
9 The Abrasive Wheels Regulations 1970.

This is not an exhaustive list

The following comments use ordinary language to convey the essence of the legislation, some of the finer points will necessarily be omitted. For a definitive version, consult the regulations and acts themselves.

The Health and Safety at Work Act 1974

This Act was very different to previous industrial legislation. Instead of specifying details of all the things to be done or avoided, it listed the general duties of employers and employees which would ensure the health, safety and welfare of people at work and members of the general public who could be affected by work activities.

The detail of how this would be done was then left to individual employers to work out. The Act makes it clear that responsibility for health and safety is shared by the employer and the employee.

Duties of employers. Some of the main duties of the employer are listed below. The language of the Act has been simplified for clarity. Employers must provide and maintain:

a) a safe place to work, with means of access and exit
b) safe systems of working
c) safe plant, equipment, machinery and tools
d) safe methods of using, storing, handling and transporting articles and substances
e) adequate instruction, training and supervision to ensure safe working
f) a safe and healthy environment and provision for employees welfare.

Designers, manufacturers, suppliers and installers must undertake all necessary research, testing, etc. to ensure the safety of plant and equipment.

Duties of employees. The main duties of employees are also listed, they must:

a) take care of their own health and safety
b) take care of the safety of anyone else who could be affected by their acts or omissions
c) cooperate fully with their employer to meet the safety requirements
d) do nothing which could interfere with, or misuse, anything provided to ensure health and safety.

Omissions are things which the employee should have done but did not, such as reporting defective machinery or equipment.

This Act requires the employee to participate fully in the maintenance of health and safety at work.

Enforcement of the Act. The Act is enforced by inspectors appointed by the Health and Safety Executive. Inspectors have the right to enter and inspect premises and have wide-ranging powers to aid an investigation of breaches of the safety laws, and to institute legal proceedings if necessary.

If an inspector feels that the safety laws are, or are about to be, breached he or she may serve an Improvement Notice on an employer or an employee requiring them to remedy the situation within a specific time.

If there is a risk of serious injury in the situation, the inspector may serve a Prohibition Notice. The unsafe activity must stop immediately, and the hazard be dealt with. A Deferred Prohibition Notice sets a deadline for the hazard to be removed, if this is not met then the activity must cease.

The inspector must state his or her reasons for deciding that a breach of the safety laws has occurred.

The Management of Health and Safety at Work Regulations 1992

These regulations provide a general framework for the management of the more detailed regulations which follow. Employers are required to:

a) assess the risks to the health and safety of employees and others from work activities
b) determine preventive, protective and emergency measures where there is a significant risk
c) plan, organise, control, monitor and review arrangements to ensure the management of health and safety, including emergency procedures
d) provide information, training and assessment to employees and others to ensure health and safety.

Employees must cooperate fully with all health and safety instructions and training, and must report all dangerous situations and any shortcomings in health and safety arrangements.

The Control of Substances Hazardous to Health Regulations 1988 (COSHH)

These regulations specify substances which have properties that could pose a hazard to health under certain conditions. They include toxic, corrosive and irritant substances, some micro-organisms, all types of dust when present in substantial concentration, and some cancer-causing substances (carcinogens). Some substances, such as hardwood dust, are hazardous when a certain exposure level is exceeded.

The regulations require employers to assess the risks to health when using hazardous substances and then to take action which will:

Health and safety legislation

Health and Safety at Work Act 1974

Employers provide:
- a safe workplace, work systems, work equipment, handling methods
- a reasonable environment, welfare facilities, instruction and training
- for employees and certain third parties.

Employees must:
- look after their own safety and that of others and comply actively with all safety requirements.

Management of Health and Safety at Work Regulations 1992

Employer to:
- assess risks
- devise health and safety measures
- plan, organise, control, monitor, review
- provide information, training assessment.

Control of Substances Hazardous to Health 1988

- specified substances, conditions
- prevention, control, monitoring
- protection measures
- control equipment maintenance, examination, testing
- health checks and records.

Personal Protective Equipment at Work Regulations 1992

- appropriate, suitable PPE
- maintenance, storage
- issue, instruction, training use, care
- reporting of loss, defects
- no charge to employee.

Provision and Use of Work Equipment Regulations 1992

- suitable, safe equipment
- maintenance, personnel, information
- guarding and protection against specific hazards
- controls, control systems
- maintenance procedures
- markings and warnings.

Workplace (Health, Safety and Welfare) Regulations 1992

- maintenance and cleanliness
- ventilation, temperature, space, lighting, layout, seating
- floors and traffic routes
- falls and falling objects
- glazing, windows, doors, gates
- sanitary, washing, drinking facilities
- changing, storing clothes
- resting and eating facilities.

Manual Handling Operations Regulations 1992

- task and method
- load size, type, difficulty, dangers
- task environment
- PPE constraints, problems
- operator constraints, requirements.

Other relevant legislation

The Factories Act 1961
The Abrasive Wheels Regulations 1970
The Electricity at Work Regulations 1989
The Safety Signs Regulations 1980

The Environmental Protection Act 1990
The Noise at Work Regulations 1989
The Reporting of Injuries, Disease and Dangerous Occurrences Regulations 1985

a) prevent exposure to these substances or where this is not reasonably practicable then

b) control the exposure of employees and others to these substances

c) if necessary, control exposure using personal protective equipment, ensuring its correct and proper use.

Equipment used for the control of these substances must be maintained, examined and tested at certain set intervals.

In some cases, the exposure of employees may need to be monitored, health checks made and records kept over a substantial time period.

COSHH and the machine operator. Special care must be taken when the machining process produces large amounts of dust, particularly grinding-wheel dust, dust from thermosetting polymers (Tufnol) and when grinding tungsten carbide tools. In some cases extraction equipment may be required.

The fumes given off by PTFE (polytetrafluoroethylene) when it becomes hot during machining can cause temporary illness.

Contaminated mineral oils can cause skin problems, including dermatitis. Remedies include changing the type of oil, changing work methods to avoid contact and using appropriate barrier creams or gloves to limit contact. However, the wrong type of glove may absorb oil and cause prolonged skin contact.

Contaminated oil in contact with the skin over long periods can cause cancers such as cancer of the scrotum. Overalls which become saturated with oils should be changed regularly and oil-soaked rags should not be kept in overall pockets.

The Provision and Use of Work Equipment Regulations 1992

These replace or revoke several parts of the Factories Act and the Abrasive Wheel Regulations.

These Regulations impose a general requirement that all work equipment must be suitable and safe for the tasks and environment in which it is used, and that all persons at risk are given information, instruction and training.

Figure 1.1 Grinding wheel with dust-extraction equipment.

'Work equipment' covers everything from a hammer to a milling machine. 'Use' includes repair and maintenance, cleaning, transporting and stopping and starting.

General requirements. The Regulations require that:

a) Work equipment must meet all EU health and safety requirements and must be maintained in safe condition by properly trained personnel. Equipment must be stable and adequately lit.

b) Training, instruction and information must be given to those employees who manage, supervise, use, maintain, repair, etc. work equipment so that they are aware of the risks, hazards and precautions involved in its use. This is particularly important for young people.

c) Direct physical contact with dangerous parts of a machine must be prevented or avoided by:

Figure 1.2 Machine guards.

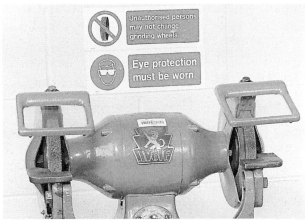

Figure 1.3 Health and safety markings.

i) stopping all movement of the part before contact is possible,

ii) preventing physical contact using guards, protection devices, push-sticks, etc., if **i)** and **ii)** are not practicable then

iii) provide information, instruction, training and supervision. This section includes a rotating stock bar protruding from a lathe headstock.

d) Guards must be suitable, effective, soundly constructed, properly sited and maintained in good order. They should allow maintenance to take place without general loss of safety, and should allow an unobstructed view of the work area if required.

e) Maintenance should be without risk if possible, else the risk must be minimised by measures such as complete machine shut down. Where direct access to dangerous areas cannot be avoided, safety systems should be used, such as permit-to-work or safety interlocks. Warning systems may be required to alert other people to dangerous situations.

f) Exposure to specific hazards must be prevented or controlled. This includes falling or flying objects, excess heat or cold, disintegration of items, fire and the uncontrolled discharge of a gas, dust or spray. Where possible, this to be done other than by personal protective equipment or training, etc.

g) Machines must have controls to isolate, start and stop, and adjust operating conditions. Controls must be visible, accessible and clearly identifiable. **Emergency stop** controls must stop a machine completely and leave it in a safe condition. Restarting must be a deliberate act. Control systems must fail-to-safety and must not affect stop or emergency stop controls.

h) Work equipment to carry all required health and safety or warning markings in a clear, visible manner. Examples are **STOP** and **START** controls, grinding wheel maximum speeds, etc.

Work equipment and the machine operator. The machine operator must be trained in the safe use of the machine and understand all the hazards involved. He or she must make proper use of all guards and protective equipment supplied and must report any safety problems immediately.

The operator should regularly check the operation of the important machine features such as stop and start controls, the emergency stop, the operation of brakes and the machine isolator.

Machine adjustment, or maintenance, which requires the removal of fixed guards must only be done by an authorised person. In some cases, it is required that the machine is locked in electrical isolation while such work is carried out; the person doing the work keeping the key. In situations where there is a high risk factor, a permit-to-work may be used. The whole operation is detailed in writing and a competent third person is used to check firstly that the work may commence safely, and then that the work has been completed and the machine may be safely used again.

The Manual Handling Operations Regulations 1992

These regulations cover the manual lifting, transport and depositing of loads.

Assessment of the risk is made considering:

a) the handling task and how it is to be done
b) the size and type of load and any inherent difficulty or danger
c) problems in the surrounding environment which could affect the task
d) problems caused by the need to wear special protective equipment which could affect the task
e) special needs or constraints on the operator such as physical strength, special training or hazards to pregnant women.

The employer must then take steps to reduce the risk of injury wherever possible, and to provide information to employees on the weight and nature of loads so that the employee will be prepared for the handling task.

Employees must make use of all equipment, training and instruction which have been given and carefully follow all laid down systems of work.

Manual handling in the machine shop. Lifting, transporting and depositing loads

Figure 1.4 Crane load with safe work load (SWL) marked.

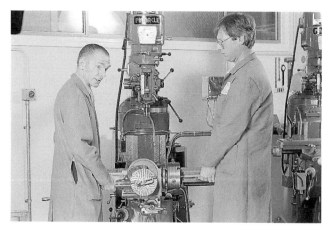

Figure 1.5 Two people lifting a dividing head.

should use procedures which do not injure the operator. Correct posture and lifting techniques are essential.

Vices, chucks, dividing heads, etc. may be so heavy or awkward that they pose a risk when handled manually. In such cases they should be marked with their weight and other relevant information.

If a two-person manual lift is to be used, both people should be of similar height and agreement must be reached on which person will direct the operation.

If possible, mechanical methods should be used such as slings, chains, trolleys, etc. Care must be taken to ensure the safety of others when using mechanical handling methods.

Plan the handling operation, prepare storage areas and ensure clear access and traffic routes.

When lifting heavy objects onto a milling machine table, adjust the knee so that the table is at a convenient height to avoid stretching up with the load.

The operator must be aware of factors causing additional hazards such as movement into or out of confined spaces and slippery surfaces due to grease, oil or coolant. If gloves are worn, they may also affect the grip of the operator on the load.

Workplace (Health, Safety and Welfare) Regulations 1992

These Regulations concern the place where work takes place, its maintenance and care, the working environment and the facilities provided.

Figure 1.6 Waste storage.

The Regulations require that:

a) the workplace, equipment, devices and systems must be maintained in a safe condition and good working order

b) the workplace must be kept clean, with floors cleaned at least once a week, and waste stored in suitable containers or removed daily

c) floors must be sound, safe, in good condition and free from obstructions and spillage

d) measures must be taken to prevent injuries from falls and falling objects. Precautions include fencing, covers and warning signs. Provision of personal protective equipment (PPE) should be a 'last resort'

e) windows should be safe to open, close, clean, be protected from breakage and clearly marked

f) doors, gates, shutters, etc. must be safe to use

g) suitable and adequate traffic routes should be provided allowing people and vehicles to move safely and without difficulty

h) there should be a reasonable environment and adequate welfare facilities.

This Regulation covers ventilation and 'fresh' air, working temperatures, lighting, space, the arrangement of the workstation and seating. It also includes toilet and washing facilities, drinking water, rest and eating areas, and the provision for changing and storing clothes.

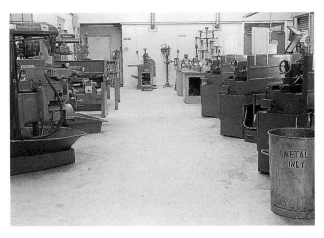

Figure 1.7 Clear gangways.

The machine operator and the workplace.
The operator must make sure that the work space around the machine is free of clutter and that he or she can move freely into or out of it.

Gangways and designated areas must be left clear. The floor should be sound and clean. Spillages of coolant and swarf should be removed without delay and stored in designated areas.

Tools and equipment should be stored safely and securely when they are not being used.

When equipment must be moved to the machine, a safe storage or seating area should be prepared beforehand.

The operator must check that there is no risk to others from his or her actions nor from the machining operations.

Overalls and personal clothes should be stored in the facilities provided. Eating and drinking must only be done in designated areas. Cleanliness is very important, hands must be washed after using toilet facilities and before eating or drinking.

Figure 1.8 Storage areas.

The Personal Protective Equipment at Work Regulations 1992

These Regulations apply to equipment designed to protect against specific hazards, such as safety glasses worn when grinding. They do not apply to ordinary working clothes. PPE is regarded as a 'last resort' when hazards cannot be controlled effectively by any other means. These regulations revoke the Protection of Eyes Regulations.

Where PPE is essential then it must be:

a) the most appropriate for the situation
b) maintained in good order and stored correctly
c) accompanied by instruction and training in its purpose, limitations, use and maintenance
d) issued without charge to the employee
e) properly used, stored and maintained by the employee who must report any loss or defects.

Figure 1.9 Eye protection when grinding.

PPE and the machine operator. If complete protection against machining grit or chips cannot be provided for the operator, then PPE may be required for eye protection. This can include safety spectacles, eyeshields or goggles. These must be appropriate for the type of hazard and must suit the operator. He or she must ensure that they are stored safely, cleaned regularly and replaced if damaged.

Consideration must be given to other people in the vicinity of the machine, including visitors.

Foot protection will only be supplied where there is significant risk of injury. Many operators buy their own safety footwear to guard against the normal hazards present in a machine shop, as well as preserving their conventional footwear from damage by swarf, oil, etc.

The Factories Act 1961

Now mainly revoked apart from sections relating to the marking and testing of lifting equipment and the employment of women and young persons.

A young person (over 16 and under 18 years of age), may not use a dangerous machine, such as a milling machine, unless fully instructed and trained in its safe operation or is under the direct supervision of a competent and experienced person.

The Abrasive Wheels Regulations 1970

Now mainly replaced by the Provision and Use of Work Equipment Regulations apart from sections relating to training and appointments. Nobody may mount abrasive wheels unless they have been properly trained and are competent. A register must be kept with the names of all people appointed to mount wheels in an establishment, together with a description of the class of wheels which may be mounted.

Figure 1.10 Milling cutter in a cloth.

General safety

- Wear overalls and maintain them in good repair, and in a clean condition.
- Make sure that no item of clothing could become entangled on rotating parts of the machine, for example shirt cuffs, ties, scarves, etc. Similarly, remove rings, watches, earrings, etc. which could be caught by swarf. Tie up long hair or wear a cap.
- Remove swarf with a swarf rake or brush, never with bare hands. Do not use brushes with looped ends.
- Wear safety shoes or boots to protect feet and toes from injury and to avoid slipping on wet floors.

- Wear safety spectacles or goggles to protect eyes. Apply barrier cream before commencing work and wash hands thoroughly before handling cigarettes, food, etc. and at the completion of work.
- Use gloves to protect hands against oils, cuts, burns, etc. Seek medical attention promptly for cuts, burns and injuries.
- Isolate the machine when it is necessary to place hands near dangerous items such as milling cutters or rotating workpieces.
- When handling sharp cutters or tools, wrap them in a dry cloth.
- Use guards and protective devices which are supplied. Make sure that guards are adjusted to provide the maximum protection against cutters, swarf, coolant, etc.

2
Metal cutting

The metal cutting process is influenced by three main factors:

a) the **cutting speed** or rate of metal removal. Too slow a speed produces a rough surface finish, too fast a speed causes rapid tool failure
b) the **tool geometry** or tool shape. This is determined by the type of cutting tool, the workpiece material and the particular machining operation taking place
c) the **feed rate** at which the tool moves through the workpiece. This is adjusted to give a balance between material removal and the desired surface finish.

Three types of chip may be formed:

a) continuous – produces long, dangerous coils of swarf. Associated with ductile materials
b) continuous with a built-up edge – causes rapid tool wear and a spiky surface on the work. Associated with high frictional and cutting forces
c) discontinuous – naturally forms small chippings. Related to brittle workpiece materials.

Chipbreaking reduces the length of chip for safety reasons and also aids chip disposal. Chipbreakers may be mechanical, ground-in or pressed-in.

Cutting fluids improve the cutting process. Fluids may be water-based, oil-based, synthetic or chemical or gaseous. Operator protection is important and disposal procedures must be followed.

Cutting metal

When the tool and workpiece are forced against each other, a compressive force is set up. The metal deforms in front of the tool point. This deformation will take place in a zone along the **shear plane**. The metal is forced to slide over the tool face and in doing this the internal structure becomes elongated and stressed.

The machined surface left on the workpiece is affected by three major factors:

a) the cutting speed or rate at which metal flows over the tool
b) the tool geometry, or tool angles
c) the feed rate.

Cutting speed

Varying the cutting speed

When the rate of metal flow over the tool is slow, the metal breaks into small segments (chips) and the machined surface is pitted and rough. The intermittent cutting process causes a fluctuating force on the tool.

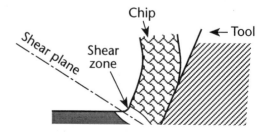

Figure 2.1 The metal cutting process.

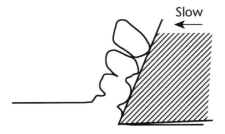

Figure 2.2 Slow cutting speed.

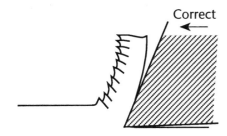

Figure 2.3 Correct cutting speed.

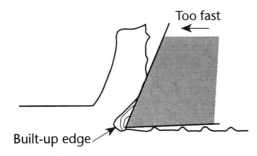

Figure 2.4 Cutting speed too fast.

If the cutting speed is increased, the segments hold together and a continuous chip is formed. The underside is shiny from sliding contact with the tool face. The topside is serrated, showing the compressed segments of metal. The workpiece surface is acceptable.

With ductile materials, the increase in friction, pressure and heat at the cutting edge causes localised welding of metal to the tool point. This is referred to as a **built-up edge**. This temporary extension of the tool alters the tool point geometry and reduces the cutting force on the tool. The built-up edge is eventually torn away and sticks to the chip and the workpiece surface. This is a continuing process and leads to rapid failure of the cutting edge. The workpiece surface is rough. This effect may be remedied by decreasing the cutting speed.

High cutting speeds produce complex internal metal flows within the chip, as the high pressure and friction reduce the speed of chip flow across the tool face. The chip is smooth, but not shiny on the underside. The workpiece surface is acceptable, but the tool wears quickly.

Recommended cutting speeds. This is the rate of material removal recommended by the tool manufacturer. It is given as metres of chip per minute (or feet of chip per minute). It is developed experimentally and will give a reasonable production rate for a specific tool life (usually 15 min). Exceeding the recommended cutting speed can shorten the tool life substantially.

Machine spindle speed. Cutting speed is related to the machine spindle speed in the following way

Spindle speed (N) (in rpm)

$$= \frac{\text{Cutting speed } (S) \text{ (m/min)}}{\pi \times \text{diameter of work (or tool)}}$$

$$N = \frac{S \times 1000}{\pi \times D}$$

The cutting speed is first changed to mm/min to suit the normal units of measurement. The diameter of the workpiece is used when turning; the diameter of the tool is used when milling or drilling, the diameter of the wheel when grinding.

The cutting speed varies with the workpiece material, the type of cutting tool, the cutting operation and the surface finish required. Each tool manufacturer provides tables of cutting speed recommendations.

Some very general guidelines are given in Table 2.1.

Table 2.1 Typical cutting speeds

Workpiece material	HSS tools (m/min)	Carbide tools (m/min)
Steels		
Free cutting	40	150
Low carbon	30	120
Medium carbon	20	80
High carbon	12	50
Aluminium alloys	90	360
Cast irons		
Medium grade	25	80
Chilled	10	35

Note: HSS = high-speed steel.

Examples of speed calculations.
Example 1. Turning ⌀50 mild steel with a high-speed steel (HSS) tool. Cutting speed = 30 m/min

$$\text{Spindle speed (rpm)} = \frac{30 \times 1000}{\pi \times 50} = \frac{30{,}000}{157.08} = \underline{191}$$

Transposing to find the cutting speed when the spindle speed is known:

$$\text{Cutting speed } (S) \text{ (m/min)} = \frac{N \times \pi \times D}{1000}$$

Example 2. Turning ⌀50 at 350 rpm

$$\text{cutting speed } (S) \text{ (m/min)} = \frac{350 \times \pi \times 50}{1000}$$

$$= \frac{54{,}977.87}{1000} = \underline{55}$$

For imperial calculation

spindle speed (N) (rpm)

$$= \frac{\text{cutting speed } (S) \text{ (ft/min)}}{\pi \times \text{diameter of work}}$$

$$N = \frac{S \times 12 \text{ (in/min)}}{\pi \times D}$$

transposing

$$\text{Cutting speed } (S) \text{ (m/min)} = \frac{N \times \pi \times D}{12}$$

The effect of tool geometry

The following examples use single-point turning tools for simplicity. Milling and drilling tools are similar in principle, but grinding wheels are a special case.

Rake angle

The rake angle is the angle between the top face of the tool and the tool base, assuming that the tool is horizontal. Note that if the tool is not on centre, or is tilted, the working rake angle will be altered. Rake angles may be classed as positive, negative or neutral. A neutral rake angle may also be called a zero rake angle.

The effect of altering rake angle.

Positive rake (see Fig. 2.5):

- area under shear decreases
- force on tool decreases
- chip thins
- frictional force decreases
- chip motion speeds up.

Negative rake (see Fig. 2.6):

- area under shear increases
- force on tool increases
- chip thickens
- frictional force increases
- chip motion slows

While a positive rake angle reduces the force on the cutting tool, there are disadvantages when it is used with carbide or ceramic inserts. The cutting force causes a bending action at the tool point, leading to premature fracture of the tool.

In addition, clearance angles must be ground on the insert, further weakening the cutting edge and leaving only the top surface of the insert for cutting purposes.

A negative rake toolholder redirects the cutting force into the body of the tool. This produces a compressive force on the insert, which is now well supported by the toolholder. The insert does not require clearance angles, so the cutting edge is stronger and both top and bottom surfaces of the insert are available for machining purposes.

Negative rake cutting increases the force on the tool by 10–15%. This can be reduced by clever design of the insert geometry. A chipbreaker groove just behind the cutting edge (land) can effectively convert the negative rake of the toolholder into a positive rake on the insert. The cutting force is reduced and a better chip flow achieved. This form is pressed into the insert when it is sintered (see Fig. 2.7).

The blunt edge presented to the work in negative rake cutting requires higher cutting speeds and more power (about four times that for positive rake cutting). Friction generated by this type of cutting helps plasticise the chip and reduce the force on the tool. The heat produced should be removed along with the chip leaving the workpiece relatively cool to the touch.

Figure 2.5 Positive rake tool.

Figure 2.6 Negative rake tool.

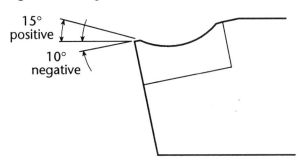

Figure 2.7 Positive rake groove.

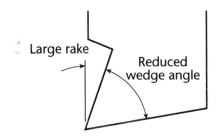

Figure 2.8 The effect of rake angle on the tool.

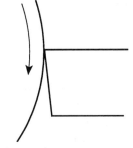

Figure 2.9 Zero rake tool.

Material structure and rake angle. In general terms it is better to have as large a rake angle as possible. However, increasing the rake angle decreases the tool wedge angle and weakens the tool. Therefore, a large rake angle is used with relatively soft materials, while a smaller rake angle is used with harder or tougher materials.

Grey cast iron, leaded metals such as free-cutting steel and brass, and machinable thermosetting plastics such as Tufnol, will all produce discontinuous chips in the form of powder, short chippings or dust. For these materials a zero rake angle is used to give a strong tool form. Varying the rake angle will not improve cutting.

Clearance angles

Clearance angles are used to avoid the front of the tool rubbing the newly cut surface. Too small a clearance angle will cause marking of the workpiece and will wear away the cutting edge. Too large a clearance angle reduces the tool wedge angle and weakens the tool.

The clearance angle will vary with the type of turning operation taking place. Typical values are shown below.

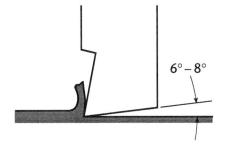

Figure 2.10 Typical clearance angles.

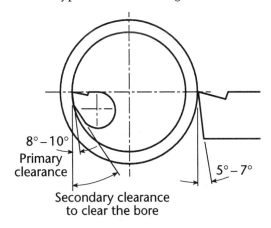

Figure 2.11

Feed rate

The feed rate is the rate at which the tool advances during one revolution of the workpiece. Generally, the slower the feed rate the better the surface finish, except that at very slow feed rates the tool may rub rather than cut, particularly with tungsten carbide tools. The feed rate is related to the tool nose radius, and should not exceed 80% of the radius value. Each tool manufacturer provides feed rate tables for different workpiece material types, required surface finishes and tool nose radius values.

As a general guide, for a good surface finish of about 1.6 μm use a feed of 0.1 mm (0.004") per revolution. For roughing operations use a feed of 0.25 mm (0.01") per revolution. These values may be used for tungsten carbide or HSS tools.

Chip formation

Continuous chip

Ductile materials like mild steel and aluminium form long continuous strings of swarf. Contributory factors include:

- a smooth, low friction cutting face
- large rake angles
- fine feeds and high speeds
- a sharp tool
- plentiful lubrication
- no chipbreaking mechanism.

Chip with built-up edge

Ductile materials form a built-up edge at the chip/tool interface when subject to high fric-

tion and pressure. This leads to rapid tool wear and a ragged surface on the workpiece. Contributory factors include:

- high friction at the tool face
- small rake angles
- coarse feeds and high speeds
- a dull tool
- inadequate lubrication with HSS.

Discontinuous chip

Brittle materials such as grey cast iron or machinable thermosetting polymers like Tufnol form powdery chips. Free-cutting materials like leaded steel or brass form small chippings when machined. In such cases, a zero rake angle is used. Coolant is not applied, however an air blast with air extraction may be used to clear the tool of chippings.

Chipbreaking

Continuous chips are dangerous and difficult to dispose of. They can mark the work, cause tool breakage and prevent coolant from reaching the cutting area. Various methods are used to break the chip into smaller disposable pieces.

Methods of chipbreaking

Mechanical chipbreakers. A hard shoulder is set behind the cutting edge to fracture the chip. The shoulder must be adjusted for different depths of cut and types of material (Fig. 2.13).

Ground-in chipbreakers. Similar to a mechanical chipbreaker, a shoulder or groove is ground behind the cutting edge. They are used on HSS or brazed carbide-tipped tools, but limit tool regrinding and carbide tips may crack when reground (Fig. 2.14).

Pressed-in chipbreakers. These are formed on carbide inserts during sintering process. They break up the chip while allowing unimpeded flow of the chippings away from the cutting area. There are many different forms (Fig. 2.15).

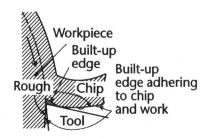

Figure 2.12 Chip with built-up edge.

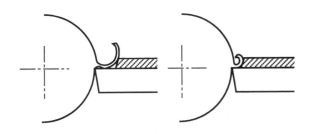

Figure 2.13 Mechanical chipbreakers.

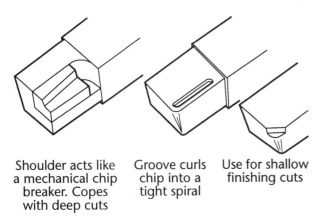

Shoulder acts like a mechanical chip breaker. Copes with deep cuts

Groove curls chip into a tight spiral

Use for shallow finishing cuts

Figure 2.14 Ground-in chipbreakers.

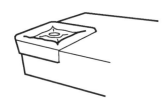

Figure 2.15 Pressed-in chipbreaker.

Cutting fluids

A cutting fluid helps the metal-cutting process by reducing friction and the effects of frictional heat at the cutting edge. It is beneficial in the following ways:

a) It cools and lubricates at the cutting edge improving surface finish, increasing tool life, allowing higher rates of metal removal and reducing power consumption.
b) It cools the workpiece reducing expansion of the part and subsequent dimensional error.
c) It washes chips away from the cutting area.
d) It helps reduce corrosion of the machined surface.

The major properties of a fluid are its ability to cool and lubricate. Other important properties are:

a) stability for a long storage and working life
b) corrosion inhibition to avoid rusting of the machine and workpiece
c) resistance to bacterial, fungicidal, chemical and physical degradation
e) safety to the operator in physical contact with it.

Types of cutting fluid

There are four main types of cutting fluid:

• water based
• oil based
• synthetic and
• gaseous.

Water-based cutting fluids

These are described as aqueous, emulsifiable or soluble oils and they form the majority of cutting fluids. The oil is added to water and it disperses as tiny globules. The water takes on a milky appearance. The oil provides lubricating properties while the water gives cooling properties to the fluid. The proportion of oil to water can be varied to cater for different machining applications.

• 1 part oil to 5 parts water for machining tough or 'difficult to cut' materials like some stainless steels
• 1 part oil to 10 parts water for heavy machining operations on the normal range of steels
• 1 part oil to 20 parts water for light machining operations on the normal range of steels
• 1 part oil to 50 parts water for grinding operations
• 1 part oil to 100 parts water for heavy machining of brass and aluminium.

Because of the high water content, rust inhibitors are added to the oil. For heavy machining operations the oil may contain additives to give improved lubrication at the tool edge.

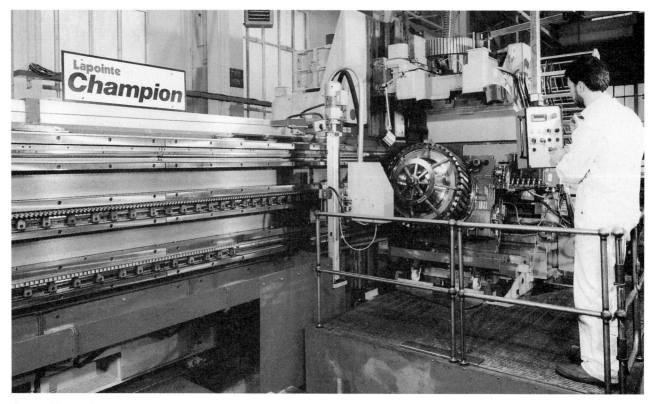

Figure 2.16 Heavy machining – broaching.

Precautions with water-based cutting fluids.

1 When mixing soluble oils always wear full protective clothing and equipment.
2 When using soluble oils always use a barrier cream or plastic gloves (where appropriate) to avoid direct skin contact.
3 Check the fluid regularly for bacterial growth, fungal growth and pH level.
4 Periodically clean out the coolant system and treat with bactericide, fungicide and corrosion inhibitors (safe to the operator) and flush out the system with cold water.
5 Dispose of used coolant through a licensed disposal company. **Never dispose of used fluids by any other means**.
6 Mop up or wipe up fluid spillages immediately. Dispose of soiled materials through a licensed disposal company.
7 **Never** use water-based cutting fluids when machining magnesium. This material may catch fire spontaneously when in the form of thin chips and can react with water to intensify the fire.

Oil-based cutting fluids

These are generally based on mineral oils. Fatty oils such as lard were widely used once but now have only limited applications because bacterial degradation can cause skin problems to operators. Oil-based cutting fluids are used where both cooling and lubricating properties are required. Additives may be introduced to improve the fluid's performance in difficult cutting conditions. Typical oil-based fluids are described below.

Straight mineral oil. A 'thin' or low viscosity oil for light cutting operations on non-ferrous and free-cutting metals such as aluminium, brass, magnesium and free-cutting steel.

Sulphurised mineral oils. These contain up to 0.8% sulphur which forms a metallic lubricating film on the tool cutting edge. This metallic lubricating film is more durable than a normal fluid film and enables the tool to cope with higher cutting forces. This cutting fluid is used when machining tough, ductile

Figure 2.17 Gear cutting.

metals particularly when drilling and reaming. Sulphurised oils can cause discoloration or even corrosion of copper and brass materials.

Sulphochlorinated and chlorinated mineral oils. Chlorine is added, with or without sulphur, to give a further improvement to the lubrication of the cutting edge. These oils are used for machining operations where there is extreme pressure on the tool and they may be described as EP oils. They are used for machining tough plain carbon steels, for chrome–nickel alloy steels and for difficult machining operations such as broaching, tapping and screw cutting.

Precautions with oil-based cutting fluids.
1 When using oil-based cutting fluids always use a barrier cream or wear gloves (if appropriate) to avoid direct skin contact.
2 Always check that the oil is appropriate for the machining operation.
3 Finely dispersed oil spray caused by some machining operations should be drawn away from the operator by extraction equipment.
4 If mineral oils are to be reclaimed for reuse, they should be sterilised and filtered before being returned to the machine.
5 Dispose of unwanted oils through a licensed disposal company. **Never dispose of these cutting fluids by any other means.**
6 Mop up or wipe up oil spillages immediately. Dispose of soiled materials through a licensed disposal company.

Synthetic cutting fluids

These cutting fluids are chemical solutions in a water base. The water gives the fluid excellent cooling properties and the chemical solution is arranged to provide the required degree of lubrication and wetting. Corrosion inhibitors are added to reduce rusting of machine parts and workpieces.

They have a long working life, are less prone to bacterial growth, are non-inflammable and are non-toxic. However, they are more expensive than other cutting fluids and many are not suitable for some non-ferrous metals such as magnesium, zinc and cadmium alloys. The simplest fluids have no lubricants or wetting agents and are most suitable for light/medium grinding operations. They have a colouring agent added to distinguish them from pure water.

Those fluids which are intended for more severe cutting operations have lubricants and wetting agents added. They are available in various forms for a wide range of machining operations including broaching and gear cutting. Parts machined with these fluids are not left with an oil film and may not require degreasing. They tend to foam which makes them unsuitable for some grinding operations.

Precautions with synthetic or chemical coolants. The precautions listed for operator protection and for the disposal of the other types of cutting fluids can also be applied to these fluids.

Gases as cutting fluids

Compressed air, carbon dioxide and liquid argon and nitrogen have all been used as cutting fluids in gaseous form. The stream of gas disperses heat and chips from the cutting zone while providing a clear view of the cutting process. There is less chance of contamination of the workpiece, although compressed air will normally be both wet and oily.

This type of cutting fluid is most suitable where wet cutting is not recommended. Examples include machining of grey cast iron and thermosetting plastics where a wet fluid would cause the chips to clog the cutting area or form an abrasive paste around the tool.

Precautions with gaseous cutting fluids.
1 Care must be taken to protect the operator from dust and chips carried by the gas. Extraction equipment will normally be required.
2 The operator will need to use a barrier cream or plastic gloves (if appropriate) to avoid direct contact with contaminants carried by the cutting fluid.
3 The operator must not be exposed to the extreme cold of gases in liquid form.
4 Gases must be serviced to remove as much contaminant as possible before being introduced into the machining area.

Cutting fluids and carbide tools. Carbide tools are prone to cracking if rapid heating and cooling takes place at the cutting edge. This can happen if a flow of cutting fluid is intermittent, is suddenly stopped or interrupted. It is essential therefore, that if a cutting fluid is used with carbide tooling, it must 'immerse' the cutting edge. This requires a large volume of fluid to be delivered to the cutting area and will require a large store of coolant to ensure a steady flow.

In practice, carbide tools are often used without cutting fluids, not only because of the danger of cracking the tip but because heat generation at the cutting edge will plasticise the chip and aid the cutting process (see page 12).

3

The lathe

The basic features of the lathe are:

a) a headstock for workholding and workpiece rotation
b) a carriage for toolholding and tool traverse
c) a tailstock for additional workholding or toolholding and tool traverse
d) guideways to ensure accurate alignment between the tool and workpiece
e) a machine body to hold all the features together
f) a power source with transmission and safety features.

The **centre lathe** is a flexible, manually operated machine for a wide range of turning operations. The **capstan lathe** is suitable for high-volume production of relatively simple shapes. The **turret lathe** is used for heavy, complex workpieces requiring many tools. The **CNC lathe** is best for complex or very accurate parts in relatively small numbers.

Introduction

The lathe is a machine tool which enables a rigidly held cutting tool to be guided past, or through, a rotating workpiece in order to produce the desired form. The basic requirements for a lathe are:

a) location and securing of workholding devices such as chucks and face plates
b) rotation of the workpiece with a selection of speeds
c) location and rigid support for various cutting tools
d) guidance and movement of the cutting tool along the required path.

These elements are contained within a machine body which is usually made from grey cast iron. Cast iron absorbs vibration and, complex shapes can be produced relatively cheaply.

Larger machines may be made as welded fabrications or with concrete bases. Modern materials, using a resin base, have been developed to give better properties of vibration damping and other effects.

Power is supplied from an electric motor contained within the machine. Provision must be made for stopping and starting the motor, for electrical isolation of the machine, for 'fail to safety' in the event of power interruption and for an emergency stop mechanism.

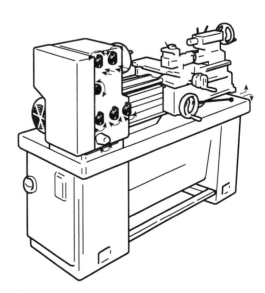

Figure 3.1 The lathe.

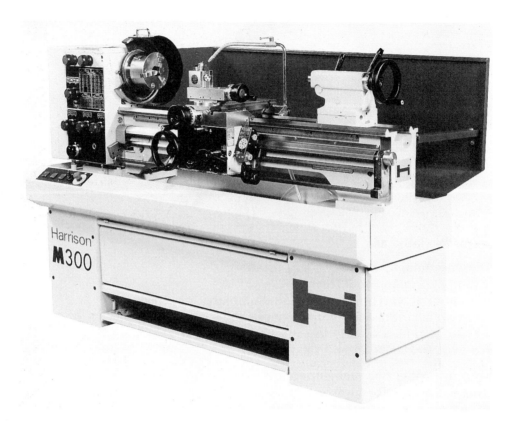

Figure 3.2 Harrison centre lathe.

The centre lathe

This is the most common turning machine and is supplied in a range of sizes to suit the workpiece range. Centre lathes are described in terms of the maximum length and diameter (swing) of workpiece that can be accommodated. Gap bed lathes allow part of the machine bed to be removed so that larger diameter workpieces can be held.

The centre lathe can be used for a wide range of machining operations making it a very versatile machine tool. It requires high levels of operator skill and is mainly used for relatively small volume work.

The headstock

This provides a central spindle on to which a range of workholding devices can be mounted. The spindle has an external taper for location and a screw thread or camlocks to secure the workholding.

There is also an internal morse taper to hold and locate a centre. The spindle is bored through its centre to hold long lengths of bar stock, which should not protrude outside of the headstock for safety reasons.

The headstock contains two gearboxes. The main gearbox enables the spindle to be driven over a range of speeds, with a larger selection of fast speeds and a smaller selection of slow speeds.

The second gearbox enables the carriage and cross-slide to be driven under power. Power is supplied by an electric motor connected to the gearbox by endless vee belts.

The carriage

The carriage holds and guides the cutting tool. The tool is rigidly clamped in a toolpost which itself is mounted on a compound slide. The compound slide is manually operated and allows the tool to be moved at an angle to the machine axis for the production of taper forms.

The toolpost and compound slide are both mounted on the cross-slide which enables the tool to be moved square to the machine axis for the production of faces, shoulders, groves, etc.

The movement of both slides is controlled by a very accurate screw-and-nut mechanism and the distance moved is indicated on graduated dials.

The cross-slide is mounted in the saddle which lies across the top of the machine bed and locates in the machine guideways.

The apron is the front part of the carriage and carries the controls for powered movement of the carriage. The carriage moves to and from the headstock, either manually, using a rack-and-pinion mechanism or by a powered traverse. This enables the tool to produce long diameters.

The carriage may carry additional tooling or workholding devices such as a travelling steady.

Figure 3.3 Compound slide

Automatic feeds. The carriage may be moved under power by engaging it with the feed shaft, for general turning, or with the leadscrew, for screwcutting operations. Both the feed shaft and leadscrew are driven by gears from the main spindle to ensure that when the spindle speed is altered, the carriage movement remains at the set feed rate.

The feed shaft usually transmits motion by bevel gears in the apron and can be connected to the carriage or the cross-slide. The leadscrew transmits motion by a split-nut to reduce backlash. A range of feed rates and screw pitches may be selected from the feed gear box in the headstock.

Figure 3.4 The lathe apron.

The tailstock

The tailstock has its own set of machine guideways. It can be moved manually along the bed and clamped rigidly in position. The barrel of the tailstock has a morse taper for a centre or for cutting tools and can be fed manually back and forwards within the tailstock body. The tailstock can be offset from the central axis of the machine to allow accurate tapers to be turned.

The guideways

Guideways are provided along the top of the machine bed to ensure alignment between the headstock spindle, the tailstock and the carriage. These guideways are generally machined

Figure 3.5 The tailstock.

into the cast iron bed and then hardened locally by induction or flame hardening before final grinding. A vee-flat form is commonly used because it allows for some wear to take place while still maintaining alignment accuracy.

Larger machines may use replaceable flat guides because of their greater bearing surface. Substantial wear of the guideways will make it impossible to produce the desired workpiece form.

The alignment of the machine features and the condition of the guideways should be checked and calibrated periodically.

Alignment checks

There are a range of standard alignment checks for lathes, covering all the major features. A simpler set of checks can be used more regularly by the lathe operator.

Headstock and tailstock alignment

Method 1. Lock the tailstock on the bed, extend and lock the barrel. Mount a dial test indicator from the spindle with the plunger on the barrel. Rotate the spindle about the barrel to detect offset from the machine axis. Adjust the tailstock position if required.

Method 2. Set a parallel test bar between centres. Mount a dial test indicator on the carriage with the plunger on the test bar, square and horizontal. Traverse the carriage and indicator to detect offset from the machine axis. Adjust the tailstock position if required.

Method 3. Set a bar of round stock between centres and turn a clean diameter for the maximum possible length. Measure the machined diameter at several places along its length. Taper of the bar can indicate offset from the machine axis.

Adjust the tailstock position if required, until an acceptably parallel bar can be turned.

Variations in diameter of a machined part can indicate guideway wear. If the tool moves off the vertical centre height, due to wear, the turned diameter will increase.

The capstan lathe

The capstan lathe is similar in principle to the centre lathe but the tailstock is replaced by a hexagonal rotary turret. Each of the turret's six faces is able to hold a number of cutting tools. The turret runs on its own special guideways, separate to those for the carriage. It is fed back and forwards either manually or by power traverse. The maximum forward position is controlled by stops. These are adjustable using special studs and there is one stop for each tooling position.

There are two toolposts. On the operator's side there is a four-way toolpost for the normal range of turning tools. At the rear of the machine there is a single toolpost used for parting-off or grooving. The tool is mounted upside-down to avoid having to reverse the chuck.

The workpiece is fed through the chuck in bar form (usually) and gripped by fast-acting collet chucks. These are round, square or hexagonal to suit the stock bar. Automatic bar-feeding equipment may be used to give rapid advance of the stock into the working position.

Capstan lathes require time-consuming, accurate setting up of tools and attachments. This is only worthwhile for long production runs using standard bar shapes. They are used for rapid production of high volume work.

The turret lathe

The turret lathe is almost identical to the capstan lathe. It is used for heavier, larger workpieces or those which require a variety of tooling.

The turret is much larger than the capstan turret and it runs on the same guideways as the carriage.

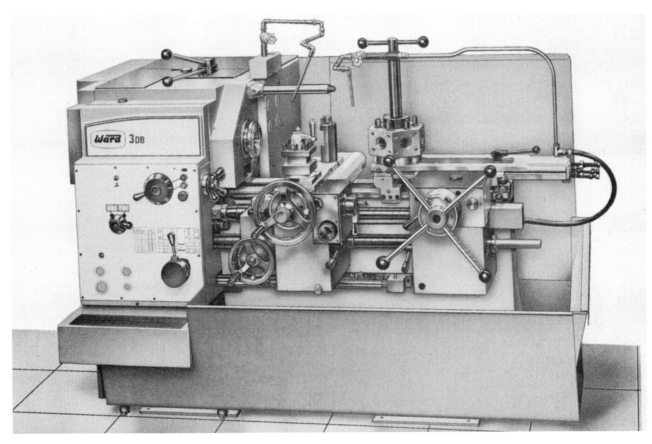

Figure 3.6 The capstan lathe.

Figure 3.7 The turret lathe.

The CNC lathe

The Computer Numerically Controlled or CNC lathe has the same basic requirements as other lathes, but the toolpath is controlled by a computer working to a computer program. The program lists a sequence of points in space which the tool must visit in order to produce the required form. These tool positions are usually referenced to fixed machine datums, similar to the process of marking out from datum faces on a component.

The machine has no handwheels or graduated dials, control of the slides is by servo-motor drives and the slides positions are monitored by devices called transducers.

The slides are moved by special types of leadscrew called ballscrews. These have virtually no backlash and very low friction, providing high accuracy and fast response.

CNC machines remove metal in larger volumes and for longer periods than manual machines. The increased volume of chips may require conveyors for removal. These lathes are usually constructed with slant beds, rather than horizontal beds, to help disposal of the chips.

CNC lathes may have one or more turrets of tools, there may be driven (live) tools such as milling cutters to avoid second operation work, and there may be automatic tool changing facilities.

The automatic operation of these machines, together, with the larger volumes of chips and coolant, require totally enclosing guards for the protection of the operators.

CNC turning reduces costs of production in the following ways:

a) repetition of the machining cycle without loss of accuracy due to operator fatigue or boredom
b) consistent and high accuracy of form
c) reduced scrap and improved quality
d) improved production rate due to reduced set-up time, more efficient machine operation and reduced use of ancillary equipment such as dividing heads.

CNC lathes are used for complex, high-accuracy parts in small volumes.

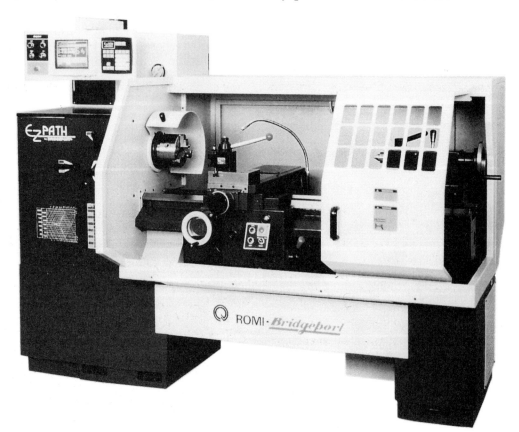

Figure 3.8 The CNC lathe.

4
Cutting tools

There are four cutting tool **materials** in general use as lathe tools: high speed steel, stellite, cemented carbides and ceramics. Each material is associated with particular **types** of lathe tool: solid toolbits, solid single-point butt-welded tools, tipped single-point tools, deposit tipped single-point tools, toolholder indexable insert tools and inserted tooth tools.

Tool **shapes** are selected to suit the particular turning operation.

High speed steel tools may be ground to any shape by the operator, but there is a selection of recommended shapes. Tungsten carbide tips and inserts are available in a limited range of shapes and sizes, but their composition may be varied. Tools can be supplied as right hand, left hand or neutral.

The most important **tool angles** are the approach angle, the major and minor cutting edge angles, the normal rake and clearance angles. Tool angles only alter with wear or regrinding. **Working angles** are those formed when the tool is set to the workpiece. Working angles can be altered by varying the tool setting. The working approach angle is set as orthogonal (square) or oblique (angled), depending on the operation. If the depth of cut and feed rate stay the same, then the chip area is constant whatever the approach angle used.

Cutting tool materials

High speed steel (HSS)

An alloy steel which retains its ability to cut metal even at elevated (700°C) temperatures (called 'hot hardness'). There are different grades of HSS. The best known is sometimes referred to as 18:4:1 to indicate the alloying elements of 18% tungsten, 4% chromium, 1% vanadium plus 0.75% carbon.

HSS can be softened by annealing and then machined into complex form tools or standard cutting tools such as drills, reamers, slot drills, end mills. After hardening, the tempering process further increases the tool hardness (secondary hardening). This metal has a hardness of 63–65 Rc and is also very tough making it good for interrupted cutting operations. It

can be supplied as solid blanks, tool bits or butt-welded tools.

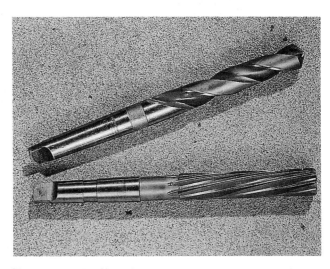

Figure 4.1 Drill and reamer.

Stellite

An expensive, non-ferrous alloy of cobalt chromium and tungsten. It is not as hard as HSS at room temperature, but it retains its hardness at higher temperatures and has a higher tensile strength. It cannot be heat-treated, rolled or forged. It is either cast into tips and brazed on a tool shank or deposited by welding on a steel shank and ground to shape.

Cemented carbides

In their simplest form, these are expensive alloys of tungsten carbide, cobalt and other elements. Carbide is harder than HSS and remains hard at up to 750°C, enabling higher cutting speeds to be used. Carbide is more brittle than HSS and needs to be well supported. It is produced as sintered tips and inserts, which are brazed or clamped on the toolholder. Although carbides can be ground with silicon carbide (green wheel) and diamond grinding wheels, they are usually disposed of when worn. If regrinding does take place, dust-extraction equipment may be required because of the danger to health from inhalation of this heavy metal.

Ceramics

The basic form is sintered aluminium oxide. They are very hard and can cut at up to 1000°C, with cutting speeds of up to 5000 m/min. The material is brittle and needs rigid support and vibration-free machines.

Cermets are ceramics in a binder of metal and carbides. They are stronger than plain ceramics, less affected by vibration, can machine a wide range of materials and can stand interrupted cuts. There are two main types, an aluminium oxide base and a silicon nitride base (Sialon). They are generally used as inserts but may be glued in place using epoxy resin.

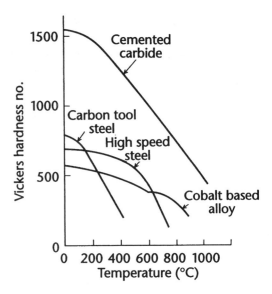

Figure 4.2 Tool materials: how hardness changes with temperature.

Types of lathe tool

Solid toolbit (HSS)

A rectangular length of HSS (toolbit) is held in a toolholder. The toolholder inclines the toolbit at a fixed rake angle. The operator must grind on the clearance angle, the tool form and any alteration to the rake angle. This arrangement lacks rigidity and only light cuts may be taken. Regrinding is a skilled job and may be severely limited by the direction of the rake angle. After regrinding, the tool must be reset to the centre of the work.

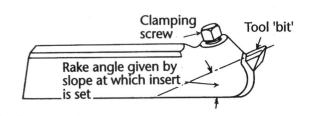

Figure 4.3 Solid toolbit.

Solid single point butt-welded tool

A piece of HSS is butt-welded to a medium carbon steel shank. This gives increased rigidity to the tool and allows higher rates of metal removal.

The tools are supplied in the standard range of shapes, with rake and clearance angles already ground in. The geometry may be altered by the operator if required.

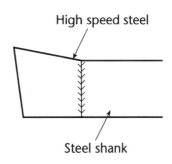

Figure 4.4 Butt-welded tool.

Tipped single-point tool

Carbide or ceramic tips are brazed to a medium carbon shank, giving a rigid tool. The tip seating imparts a fixed rake to the tool, clearance angles are dictated by the tip form. A variety of tip shapes are available. Limited regrinding is possible, but this can cause thermal stresses and cracking of the tip. Grinding in a chipbreaker form also limits possible regrinding.

Figure 4.5 Tipped tool.

Deposit-tipped single-point tool

Hard Stellite is deposited on a carbon steel shank by welding. This is ground to the required geometry using an aluminium oxide wheel. This provides a rigid tool with superior abrasion resistance, which can be run at high cutting speeds. The increased speed requires better support at the tool edge so a smaller clearance angle is used. They are rather expensive in comparison to the more flexible cemented carbides. They are used to cut cast iron at high speeds.

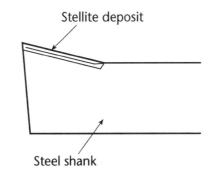

Figure 4.6 Deposit-tipped tool.

Toolholder indexable inserts

Carbide or ceramic inserts are clamped on a medium carbon steel shank, giving a rigid cutting tool. Rake and clearance angles are preselected, with negative rake angles used to give better support to the insert. A limited range of shapes are available to cover all turning needs.

When the insert is worn it is indexed to present a new cutting edge. Resetting of the tool height may be avoided by using close tolerance inserts. When all the edges are worn, the insert is discarded. The cost of these inserts is recovered by reduced changeover time and the better accuracy and consistency derived from using standardised cutting tools.

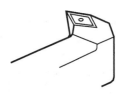

Figure 4.7 Indexable insert tool.

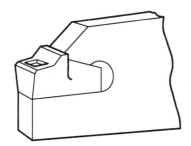

Figure 4.8 Inserted tooth tool.

Inserted tooth cutters

Specialised tools contain carbide tips inserted into the body of the tool. This gives a very rigid arrangement. The example shows an inserted tooth parting off tool.

Turning tool shapes

Tools may be right hand, left hand or neutral. This 'handing' of the tool refers to the position of the cutting part in relationship to the shank. A right-hand tool has the cutting edge on the right when looking down at the cutting face, with the shank pointing away. A left-hand tool has the cutting edge on the left. A neutral tool either has cutting edges to both right and left or else has a cutting edge directly aligned with the shank.

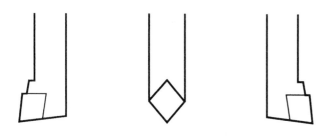

Figure 4.9 Left-hand, neutral and right-hand turning tools.

Single-point HSS tools

Light turning and facing tool

This tool can be used for facing and longitudinal turning. It can reach into awkward corners. Used for light or finishing cuts only. Saves toolchanging.

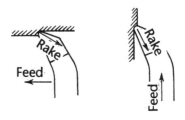

Figure 4.10 Light turning and facing tool.

Straight-nosed roughing tool

This tool is used for heavy roughing cuts. The increased approach angle makes it more efficient than the alternative round-nosed tool.

Figure 4.11 Straight-nosed roughing tool.

Round-nosed tool

This tool is used for heavy roughing cuts. The straight-nosed rougher is generally preferred.

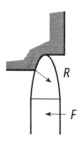

Figure 4.12 Round-nosed tools.

Knife (side cutting) tool

This tool is used for facing end faces, surfacing to corners and facing shoulders.

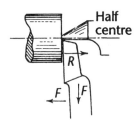

Figure 4.13 Knife tool.

Parting-off tool

This tool is used for grooving as well as parting-off to length. Normal cutting speeds may be reduced by up to 50%. Ensure that the tool is sharp, on centre height and square to the machine axis. Adjust the tool to give the minimum extension out of its holder. Lock the saddle and compound slide to avoid side play. The parting-off blade width is related to the depth of cut required.

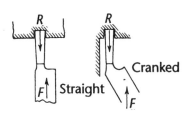

Figure 4.14 Parting-off tool.

Recessing tool

This is used to cut grooves in faces. This tool is similar to the parting-off tool, but additional side clearances will be required to clear the newly cut recess.

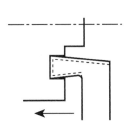

Figure 4.15 Recessing tool.

Bar-turning tool

This is a robust tool, used to take heavy cuts in black (scaled) bar. Use deep cuts, moderate to slow feeds and ample coolant.

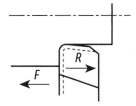

Figure 4.16 Bar-turning tool.

Screw-cutting tool

Forms the screw thread profile while generating the helix. This requires a very slow speed so that the tool may be withdrawn manually at the end of the cut. This may cause a poor finish on the thread. Use chaser tools to improve the finish. The tool must be set square to the workpiece using a setting gauge. Various infeed methods are used (see page 53).

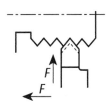

Figure 4.17 Screw-cutting tool.

Knurling tool

This tool deforms rather than cuts the workpiece surface, to produce a serrated form. Knurls may be straight or angled. Use slow speeds and feeds and ample coolant. A bright bar may require turning to remove the stressed outer skin. Older knurls may be angled so that one corner does the forming. Balanced knurls reduce the side loading on the spindle bearings caused by pressure knurls.

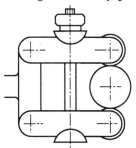

Figure 4.18 Balanced knurling tool.

Inserted tip tools

There are 16 different shapes of indexable insert available. The most widely used shapes for turning are the rhomboid and the triangular because they provide for a wide range of operations. Other shapes have specialised applications.

Square inserts

Square inserts are the strongest and can give eight cutting edges, but they cannot be used to turn external diameters and shoulders at one setting.

Rhomboid inserts

Rhomboid inserts are strong, can give four cutting edges, and allow turning of diameters and faces. There can be difficulty in machining near centres.

Rhomboid – 80° included angle. This is the strongest of the rhomboid shapes and allows turning of faces, diameters, shoulders and outward turning of large angle chamfers.

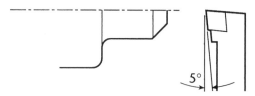

Figure 4.19 Eighty degree rhomboid.

Rhomboid – 35° included angle. This is the weakest of the rhomboid shapes and it is mainly used for finish profiling operations. As a right-handed tool it allows inward turning of angles up to 48° (35° boring) as well as all the other range of operations. As a neutral tool it allows inward and outward angles of up to 70° turning, as well as external diameters.

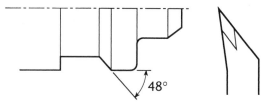

Figure 4.20 Thirty-five degree rhomboid – external right-hand.

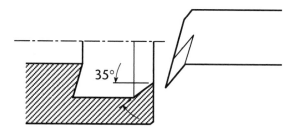

Figure 4.21 Thirty-five degree rhomboid – internal.

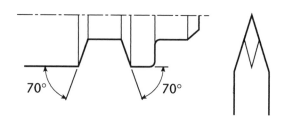

Figure 4.22 Thirty-five degree rhomboid – neutral.

Rhomboid – other angles. Other rhomboid shapes available are 55° and 75° included angle.

Triangular inserts

Triangular inserts are not as strong as square or rhomboid inserts, but they can give six cutting edges, machine diameters and faces, and allow better access when turning on centres.

Triangular – 60° included angle. The relatively small included angle allows inward turning of angles. With a right-handed tool: square faces and shoulders, external diameters, inward angles up to 25° and radii. With a neutral tool: external diameters, inward and outward angles up to 57°.

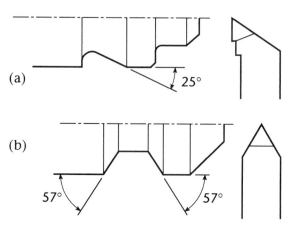

Figure 4.23 (a) Triangular – right-hand. (b) Triangular – neutral.

Round inserts

Round inserts provide the largest amount of cutting edge, but cannot machine sharp corners and can promote chatter finishes. They can be used for profiling of complex external forms.

Figure 4.24 Round.

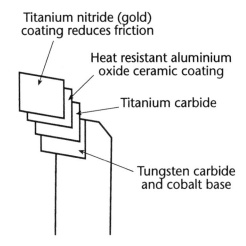

Titanium nitride (gold) coating reduces friction

Heat resistant aluminium oxide ceramic coating

Titanium carbide

Tungsten carbide and cobalt base

Figure 4.26

Screw cutting and parting-off

Specialised inserts are available for screw cutting, parting-off and grooving operations.

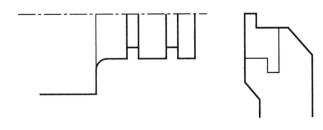

Figure 4.25 Grooving.

The carbide grade

The proportions of hard tungsten carbide and tough cobalt (the binding material) can be altered to give varying degrees of hardness and toughness. This is called the carbide grade.

Additionally, thin coatings (0.05 mm) of titanium carbide (TiC), titanium nitride (TiN) and aluminium oxide can improve tool life and the range of metals cut. These triple-coated carbides are used to rough and finish steels and non-ferrous metals. Coated carbides are not normally used for milling operations, any regrinding removes the coating.

Grade selection

Carbide tools are first selected according to the type of workpiece material being machined. The grade of carbide is identified by a letter/colour code. Then the hardness/toughness characteristics of the tool are selected according to the type of machining operation taking place. This is shown by a number between 1 and 50; 1 is very hard, 50 is very tough. This only truly applies to uncoated tools. Finally, selection may be made to take account of particular tool wear patterns when machining.

Workpiece material

• **Grade P.** Colour blue. Steel, cast steel and malleable iron producing long chips.

• **Grade M.** Colour yellow. Steel, cast steel, manganese steel, cast iron, malleable cast iron, austenitic steel, free-cutting steel.

• **Grade K.** Colour red. Cast iron, chilled cast iron, malleable iron producing short chips, hardened steel, wood, plastics and non-ferrous metals.

Number selection

Meaning of the numbers
1 Increasing
hardness ← → Increasing
toughness 50

Selection and application.

• 1 – finishing and light
  roughing

  high cutting speeds
  and light feeds

• 20 – light and medium
  roughing

• 40 – heavy roughing

  low cutting speeds
  and heavy feeds.

Tool wear patterns

Crater wear when
machining steel at high
cutting speeds
(see Fig. 4.40)

P grade resists
wear on the
cutting face. M
grade less so

Flank wear when
machining cast iron
(see Fig. 4.41)

K grade resists
wear on the flank.

Tool geometry

BS1296 (Part 2) is based on the Normal Rake System, and uses two reference systems for single-point tools, tool-in-hand and tool-in-use.

Tool-in-hand

Used when producing, sharpening or measuring tools. This system places the word **Tool** before all angles, e.g. Tool Approach Angle. The most important angles are shown here on a right-hand straight-nosed roughing tool (Fig. 4.27). The Tool Cutting Edge Angle and the Tool Approach Angle must add up to 90°. The Tool Approach Angle shown here is positive. A Tool Cutting Edge Angle greater than 90° must have a Negative Tool Approach Angle.

Tool-in-use

Angles may effectively be altered by the way that the tool is set to the workpiece. To emphasise this tool and workpiece combination, a second reference system is used. The word **Working** is put before all the angles, e.g. Working Clearance Angle.

Figures 4.28 and 4.29 show how the Working Approach Angle can be altered by the tool/workpiece set-up even though the Tool Approach Angle remains the same.

Figures 4.30–4.35 show how the angles are formed when grinding a side cutting or knife tool from a solid HSS blank or toolbit.

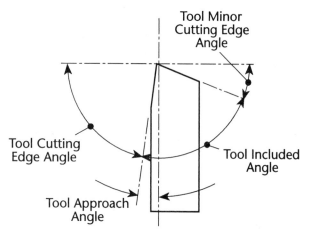

Figure 4.27 Single-point tool geometry.

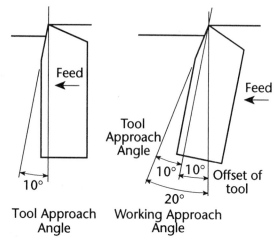

Figure 4.28   **Figure 4.29**

1 Unground toolbit

2 Grinding the Tool
   Major Cutting Edge. Responsible for stock removal.

3 Grinding the Tool Minor Cutting Edge
   Contributes to surface finish.

4 Grinding the Tool Orthogonal Clearances.
   The minimum to avoid rubbing.

5 Grinding the Tool Radius. A major factor in surface finish. Large, but not so as to cause chatter.

6 Grinding the Tool Normal Rake. Determines tool load and chip flow. Related to workpiece material.

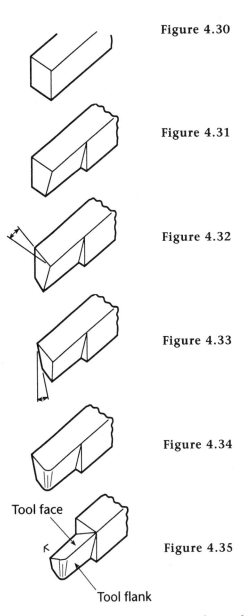

Figure 4.30

Figure 4.31

Figure 4.32

Figure 4.33

Figure 4.34

Tool face

Figure 4.35

Tool flank

Figure 4.36 shows how the Working Rake Angle and the Working Clearance Angles are affected by incorrect tool setting. The effects of these changes are listed. These effects are reversed for internal turning operations.

BS1296 (Part 2) is based on the Normal Rake System. This means that the rake angle is measured square to the Tool Major Cutting Edge and follows any inclination of this cutting edge.

Tool setting for machining

Changing the working approach angle alters the chip form and the loading on the tool.

Left: tool set below centre, external turning, reduced work rake, increased working clearance, tool may 'dig in', tool chatters.

Right: tool set above centre, external turning, increased working rake, reduced working clearance, front of tool rubs work, chatter finish.

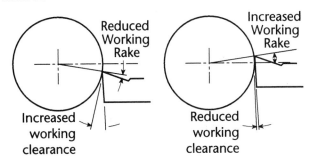

Figure 4.36 Tool setting errors.

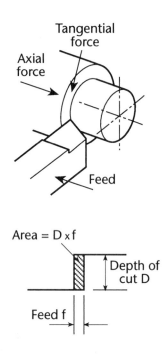

Figure 4.37 Orthogonal cutting.

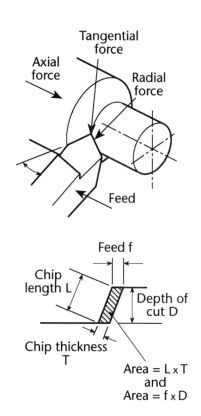

Figure 4.38 Oblique cutting.

Orthogonal cutting. A working approach angle of 0° is called orthogonal cutting. The chip area is set by the feed rate and the depth of cut. There are two forces acting on the tool, an axial or feed force and a tangential force. The tangential force is the most important and is related to the shear strength of the workpiece.

Oblique cutting. When the working approach angle is not 0°, this is called oblique cutting. The chip has the same area as in orthogonal cutting, but is longer and thinner. This spreads the force over a longer cutting edge, improving tool life. Alternatively, it can allow increased cutting speeds to be used for the same tool life, so shortening production time.

An additional force is introduced with oblique cutting: the radial force. This tends to deflect the workpiece and cause chatter, so limiting the extent of the approach angle (usually about 15° but can be extended to 45° with a rigid set-up). Oblique cutting is generally used when roughing. The cutting process is more efficient, but it is not possible to machine a square shoulder without resetting the tool. Oblique cutting also occurs with combined turning and facing tools. In this case, the feed force can push the tool into the workpiece and can result in removal of too much material from the workpiece. Care must be taken to ensure that backlash has been eliminated when setting the tool for the cut.

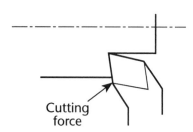

Figure 4.39 Oblique cutting with a turning and facing tool.

Tool setting and tool materials

A long, thin chip is preferred when using HSS tools. The cutting force is spread over a longer tool length and the chip stays cooler. In general, these tools are used with deep cuts and low feed rates.

When using carbide tools, heat generated by the cutting process is used to plasticise the chip, so reducing the load on the tool. A shorter, fatter chip is preferred because it retains frictional heat with its smaller surface area. As the chip leaves the cutting area it takes heat with it, leaving the workpiece relatively cool.

In general, these tools are used with deep cuts and coarse feed rates, aim for a depth of cut of 10 × feed rate as a minimum where possible.

Tool wear and tool geometry

The particular combination of depth of cut, feed rate and cutting speed produces a distinctive pattern of wear on the tool. There are two main forms of tool wear: flank wear and crater wear. If these occur in a balanced way, then tool geometry and tool life can be prolonged. If they are not balanced, the tool will fail prematurely. Tool failure is apparent when surface finish is unacceptable or when component size cannot be achieved, tool wear alone is not an indication of tool failure.

Flank wear. This takes place on the major cutting edge and the tool nose radius, producing vibration and poor surface finish. Excessive wear may be caused by too high a cutting speed or the tool set above centre. With carbide tools it can indicate that the wrong type of carbide has been selected, use a K grade.

Figure 4.40 Flank wear.

Crater wear. This takes place on the top face of the tool and can cause the tool edge to fail. Excessive crater wear indicates that the force on the tool is too high; too high a cutting speed, an incorrect feed rate or rake angle. With carbide tools it can indicate that the wrong type of carbide has been selected, use a P grade.

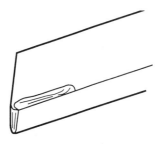

Figure 4.41 Crater wear.

Balanced wear. Flank and crater wear are such that the cutting edge is maintained.

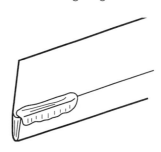

Figure 4.42 Balanced wear.

5

Lathe workholding

Workholding methods include:

- Self-centring chucks – two-jaw for premachined work, three-jaw for bright drawn bars and six-jaw for thin rings or tube.
- Independent four-jaw chucks for irregular or offset work.
- Faceplates for awkward shapes or machining to an existing datum face.
- Collets for production work on bright drawn or pre-machined parts.
- Steadies – fixed or travelling, to support long workpieces.
- Mandrels for concentric turning to existing bores or for thin or threaded parts.
- Spigots for high volume, concentric turning to existing bores.
- Centres – live or dead, for accurate, concentric turning of several diameters or for taper turning. Running centres for higher speeds, half centres when turning the end of bars.
- Bungs provide false centres in the ends of bored workpieces.

Universal chucks

Types of chuck mounting

A tapered spigot and bore provide location. Fastening may be by screws, a threaded collar or by cam-lock studs.

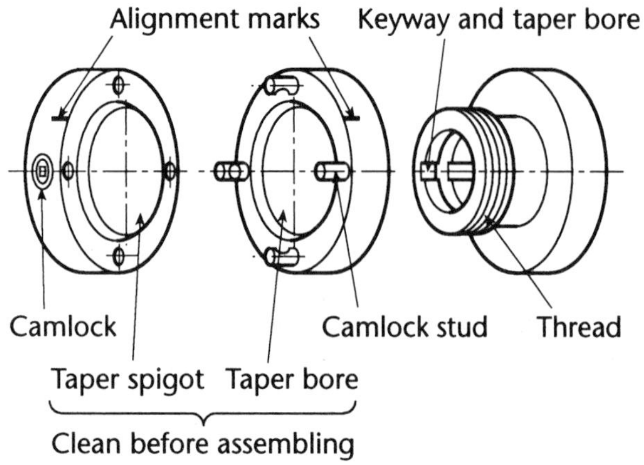

The self-centring chuck

This adjusts to suit various internal and external diameters, using two sets of hard serrated jaws. The jaws move simultaneously on a geared scroll, bringing the work concentric with the lathe axis to within 0.05 mm. Workholding is firm and rapid, but grip deteriorates with wear of the scroll.

Alignment marks   Keyway and taper bore

Camlock

Taper spigot   Taper bore

Camlock stud   Thread

Clean before assembling

Figure 5.1

Figure 5.2

Work is limited to clean, geometrically true, round and hexagonal shapes, usually pre-machined or cold-drawn bar. No attempt should be made to grip odd-shaped or scaled (black) bar.

It is difficult to reset work to run concentric once it is removed from the chuck, so complete as many operations as possible at one setting.

Thin rings or tubes are held in six-jaw chucks to spread the gripping force and avoid distortion.

The serrated jaws tend to mark the work, so where it is important to avoid this, soft jaws are used. These are attached to a special two-jaw chuck and bored or turned to suit the component size. The jaws are set to grip on a disc or ring before boring/turning to size.

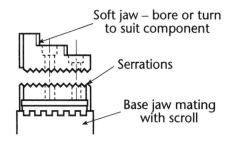

Figure 5.3 Soft jaws.

Independent four-jaw chucks

Each jaw is adjusted separately by a square-threaded screw. Large gripping forces are possible, so these chucks are bigger and heavier than the scroll-type. The jaws can be reversed on their screws so only one set is required.

Adjustment of individual jaws allows for much more accuracy when setting the work (concentricity to within 0.02 mm). Irregular shapes and scaled (black) bar can be held securely. Work can be set off-centre for eccentric boring and turning. (Balancing will be needed, see page 38.)

Figure 5.4 The four-jaws chuck.

Limited range chucks

Self-centring diaphragm chucks

These use a large spring washer to give the gripping force on self-closing jaws. The jaws are opened by an air-cylinder piston (pneu-matically). The gripping force is high, but a set of jaws can only hold a limited range of diameters once set in position. This factor, and the rapid operation of these chucks, make them most suitable for production work on pre-machined or cold-drawn bar.

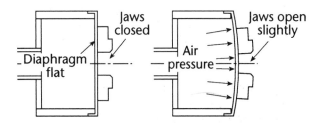

Figure 5.5 Air-operated diaphragm chuck.

Collet chucks

The chuck closes on the workpiece when it is drawn against a tapered seating. The chuck may be pulled into the seating (draw back collet), moving the work back slightly. The chuck may be pushed into the seating (push out collet), pushing the workpiece forwards slightly. If it is important that the workpiece does not move, then the collet is rigidly held against a shoulder and the seating is pushed against it (dead length collet). Individual collets can only work with a limited range of size variation (±0.05 mm) and the workpiece must be geometrically true and clean. They are supplied in a range of collet shapes (round, hexagonal, square, etc.) and sizes, and are used on production work, often on second-operations (following initial machining).

Figure 5.6 Collett chucks.

Face plates

These are mounted directly to the spindle, like a chuck, and are used to hold odd-shaped workpieces or perform operations which cannot be done in a chuck. They are particularly useful in boring or turning parallel or square to a flat surface.

Balancing work. If the mass of the workpiece is not central on the lathe axis, out-of-balance forces can arise causing vibration, damage to the spindle bearings and gears and 'chatter' marks on the workpiece. In extreme cases the work could be flung from the lathe. In such cases, the set-up must be balanced before running the machine at high speed.

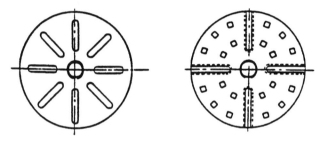

Figure 5.7 Face plates.

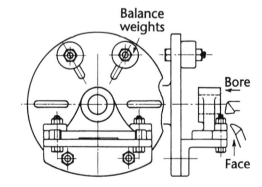

Figure 5.8 Balancing work.

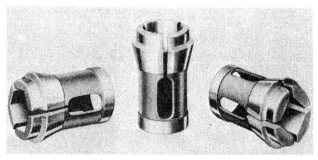

Static balancing. Place the machine in high gear or take it out of gear altogether. Rotate the spindle and mark the bottom-dead-centre (BDC) position. Repeat this procedure. If the mark consistently returns to the BDC position, this shows the area of most mass. A balancing weight must be firmly attached to the set-up directly opposite this position. Eventually the spindle will rotate freely and no one point will repeatedly return to the BDC position. This indicates an acceptable degree of balance.

Steadies

The fixed steady

This is rigidly attached to the machine bed and provides three-point support to the workpiece. It is used when the workpiece is very long and would tend to sag under its own weight or deflect away from the cutting tool. The tool cannot traverse past this type of steady. It is also used to support the end of shafts when the tailstock is being used to drill, ream, bore, etc. the shaft end.

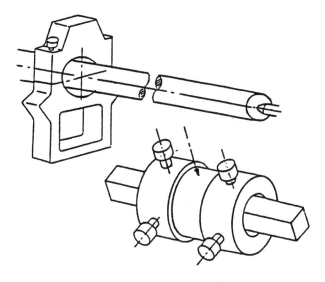

Figure 5.10 The cathead.

A clean diameter is carefully turned on the shaft where the steady is to run, and the jaws adjusted so they just touch the bar. A dial test indicator may be used to check that the shaft is running true along its length (top and side). The jaws must be lubricated while the shaft is turning.

If it is not possible to turn a true 'spot', then a **cathead** may be used. It must be carefully adjusted to run perfectly true to the machine axis.

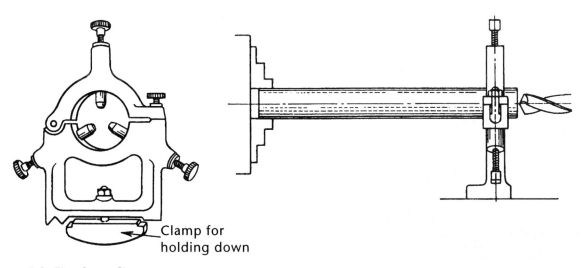

Clamp for holding down

Figure 5.9 Fixed steadies.

The travelling steady

This is attached rigidly to the carriage and travels with the cutting tool, providing two-point support to the work as it is being machined. It stops the workpiece bending away from the tool and ensures a constant diameter along the shaft length. It is used when turning long and slender shafts or when screw-cutting a long bar.

With the shaft supported on a centre, a clean diameter is turned at the end of the shaft and the jaws adjusted so that they just touch the bar, behind the tool. The jaws must be lubricated during the turning operation.

Without steady support, a barrel-shaped shaft may be produced. If the jaws are incorrectly adjusted with too much pressure, a concave shaft may be produced.

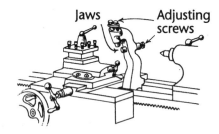

Figure 5.11 The travelling steady.

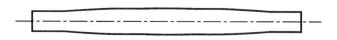

Figure 5.12 Without steady support.

Figure 5.13 Incorrectly adjusted steady.

Mandrels

A mandrel is used when it is necessary to machine external diameters dead concentric to a bore. Using a standard three- or four-jaw chuck, the hole is drilled, bored to true up the hole, then reamed for finish and size. The workpiece is then driven on a mandrel, being held in place by friction or by a nut, washer and thread. The mandrel is mounted between centres and the outside diameters turned to size. Machining takes place towards the chuck.

The plain or standard mandrel

The plain mandrel is hardened and ground with a taper of about 0.05 mm over a 150-mm length. The nominal diameter is about one-third the way along the mandrel length. Protected centres are machined both ends and a flat provided for the carrier.

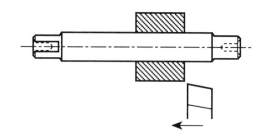

Figure 5.14 Plain mandrel.

The gang mandrel

The gang mandrel is used when machining several workpieces at the same time, particularly thin parts. The mandrel has a parallel diameter for location and a nut and washer to hold the work in place.

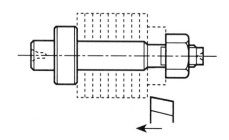

Figure 5.15 Gang mandrel.

The threaded mandrel

This mandrel is used with nuts or internally threaded parts. The workpiece must have one face dead square to the thread axis, and this is tightened against the mandrel shoulder.

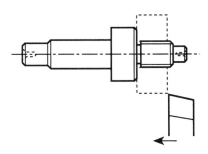

Figure 5.16 Threaded mandrel.

The expanding mandrel

The expanding mandrel has blades which adjust radially, on a taper, to work with a limited range of diameters. It must not be expanded beyond its stated limit of size or distortion may occur.

The taper shank mandrel

Instead of running between centres, these mandrels are mounted directly into the headstock taper. The method of workpiece mounting can be any of those described above.

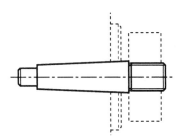

Figure 5.17 Taper shank mandrel.

Spigots

A spigot may be used instead of a taper mandrel when turning outside diameters concentric to an existing bore. It is suitable for large quantities or where centres are not available.

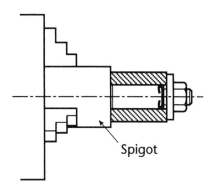

Figure 5.18

The spigot is turned in a three-jaw chuck, to provide a locating diameter and shoulder for the work. The spigot length is just shorter than the work and the end is tapped or screwed for fastening. The spigot material should be different to the work to avoid localised cold welding during machining. The spigot is lightly oiled before the workpiece is assembled and machined.

The accuracy of the spigot will be lost when it is removed from the three-jaw chuck.

Bungs

These provide false centres in the ends of a bored workpiece so that it may be machined on centres. Bungs may be parallel or tapered to fit the bore.

The bung is turned in a three-jaw chuck to provide a locating diameter or taper for the workpiece bore. It is machined with a left-hand tool so that the centre can be produced at the same time. The bung material should be different to the workpiece to avoid localised cold welding during machining. The bung is driven into the workpiece and mounted on centres for machining.

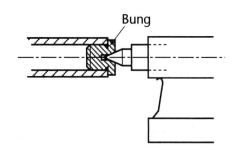

Figure 5.19

Centres

These support workpieces for concentric or taper turning of diameters.

Dead centres

The dead centre is static in its housing and is held in place by friction, usually on a morse taper shank. Dead centres normally have an included angle of 60° but larger angles (75° or 90°) may be used when the machining forces are very high. Dead centres provide rigid support and promote good surface finishes. They are greased to reduce frictional heat between the centre and the work, otherwise local welding may take place and the workpiece will lose its support during machining. They are used with HSS tools but are not suitable for the high spindle speeds required with carbide tools.

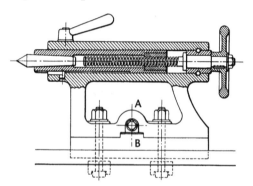

Figure 5.20

Live centres

A live centre is similar to a dead centre, but it is driven into the headstock and it rotates with the machine spindle and the workpiece.

The live centre must run dead true to the machine axis and the taper is often machined

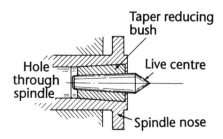

Figure 5.21

while the centre is in the headstock to ensure true running. In such cases, the centre is machined from a soft material using the compound slide and standard tools. The workpiece can bed itself into a soft centre and is better supported.

Running centres

A bearing arrangement allows the centre to rotate with the workpiece, as well as providing it with support.

Running centres are used when turning at high speeds (above 300 rpm). The clearances in the bearing give reduced radial support to the work, and vibration and chatter marks can result. This is particularly true with carbide tools. Running centres are larger than dead centres and this may cause problems when machining with standard tools. Most centres are supplied in a hardened state for increased life. Periodically they must be reground to maintain the 60° included angle.

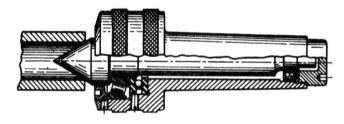

Figure 5.22

Half-centres

These are dead centres cut away to the centre-line to allow tools to face up to the centre hole. They can only be used with low speed, low force operations.

When parallel turning on centres, it is vitally important that the centres are dead in line. This can be checked by using a parallel test bar and a dial test indicator or by clocking in the tailstock spindle.

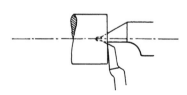

Figure 5.23

6

Turning operations

Turning produces diameters, faces and tapers by **generating** and **forming** processes. **Generating** requires a combination of tool movement and workpiece rotation to produce the finished form. Tool shape is relatively unimportant. **Forming** reproduces the tool shape in the workpiece. Screwcutting involves both generating and forming. Special operations include eccentric turning and boring, producing accurate tapers and screwcutting.

Introduction

Turning operations may be classified as either

- **generating** or
- **forming**.

Generating operations

These comprise the majority of turning operations. The workpiece shape is produced by a combination of tool movement and workpiece rotation. The tool shape has no effect on the resulting shape of the workpiece, although it may have a profound effect on the surface finish and accuracy.

There are three main generating operations:

- flat faces and shoulders
- cylindrical surfaces – external diameters and internal bores
- tapers and chamfers – internal and external.

The accuracy of generated forms depends on the machine tool construction and the accuracy with which the tool is guided.

A. Flat faces and shoulders

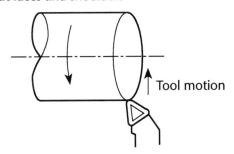

B. Cyndrical surfaces – external diameters and internal bores

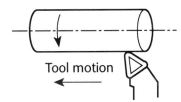

C. Tapers and chamfers – internal and external

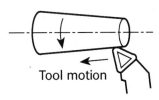

Figure 6.1 Generating processes.

Forming operations

In these operations the shape, or form, of the tool is reproduced on the workpiece when the tool is plunged into the work. Complex profiles may be produced very quickly using this method, but the long cutting edge tends to cause vibration and the surface finish may be marred by 'chatter' marks. The tool must be set on the work centre to produce a true form.

Form tools can be expensive to produce, with the form ground into the tool having to be modified to take account of the tool rake angle. The resulting profile may be very complex. The accuracy of the workpiece is entirely governed by the accuracy of the tool.

Examples of forming include large chamfers, tapers and grooves.

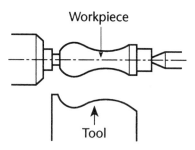

Figure 6.2 Forming.

Combined generating and forming operations

The production of a screw thread is an example of a combined operation. The screw thread profile is formed by the screw cutting tool, the screw thread helix is generated by the tool feed and the workpiece rotation.

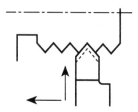

Figure 6.3 Forming and generating.

External operations

General external operations

These operations are relatively straightforward so long as the correct tool has been selected and set up, the appropriate speeds and feeds have been set up and due care is exercised by the operator. The surface finish provides a good indicator that all is well.

The unwanted metal is removed by roughing cuts, leaving about 0.5 mm for a finishing cut to give the required dimensional size and surface finish. When roughing, aim to have a single cut for each surface. Set the tool leaving the finish machining allowance, adjust the feed rate to suit the power available on the machine and set the cutting speed about 10% below the finish speed. If possible, use a cutting edge angle of 75° (approach angle 15°) to give a more robust tool arrangement. If the workholding setup is not rigid, more than one roughing cut may be required for a surface.

The finishing cut is taken with a lighter feed about 0.1–0.15 mm/rev, and at the recommended surface speed. Use a cutting edge angle 90°–95° and/or a minor cutting edge angle of 5°–10°. When turning up to, and then facing, a shoulder, use a cutting edge angle of 93°–95° (approach angle −3° to −5°).

Dimensional control

A diameter is turned on the work, measured and used as a reference for tool positioning. The tool is moved using the handwheels and the amount of movement indicated by the calibrated dials. The dials can be 'zeroed', for convenience. Most modern machines use a direct reading system on the cross-slide dial, this shows the reduction in the workpiece diameter and makes dimensional control easier. On many older machines, the dial shows the actual tool movement and the resulting reduction in workpiece diameter must be calculated using 2 × this distance.

The compound slide shows the actual tool movement because it is used for many other operations than simply turning diameters.

Longitudinal movement of the tool may be shown by a calibrated dial on the carriage. Special 'stops' can be clamped to the machine

(a)

Workpiece
reduced
by 5 mm

Tool
moves
2.5 mm

Dial reads 5 mm

(b)

Workpiece
reduced
by 5 mm

Tool
moves
2.5 mm

Dial reads 2.5 mm

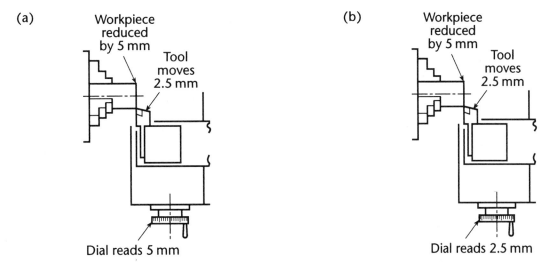

Figure 6.4 (a) Direct reading and (b) indirect reading.

bed and used to provide a fixed, adjustable datum for positioning the carriage. This makes machining to length much faster.

Using the automatic feed mechanism will improve dimensional uniformity as well as providing better surface finish.

Eccentric external turning

When two diameters must be eccentric to each other one of the following methods may be used.

Eccentric turning on centres. Sets of centres are provided on the required turning axes. The workpiece is mounted on the appropriate set of centres and the diameters are turned.

The accuracy of the part depends on the accuracy with which the centres were produced.

Eccentric turning using a four-jaw chuck. The main axis of the part is offset by a distance equal to the eccentric dimension. This is

measured using a dial test indicator and the offset is correct when the dial test indicator reading is 2 × the eccentric dimension.

Eccentric external turning to an external datum. At times it is necessary to produce a diameter which is not central on a part and which is not set to another diameter but to an external datum. In this case, the part may be set in four-jaw chuck or on a face plate using the following procedure.

The centre of the diameter is marked out using a vernier height gauge and accurately centre punched. The part is roughly set in position using a fixed centre in the tailstock as a visual guide. (The tailstock must have been accurately set on the machine axis.)

A second fixed centre is placed between the centre in the tailstock and the centre-punched centre in the workpiece.

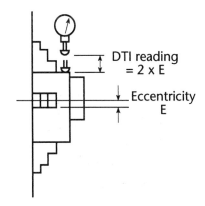

DTI reading
= 2 x E

Eccentricity
E

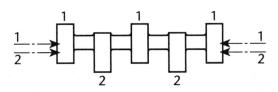

Figure 6.5

Figure 6.6

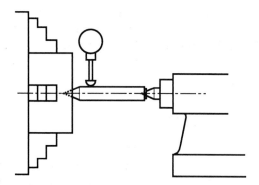

Figure 6.7 Setting to an external centre.

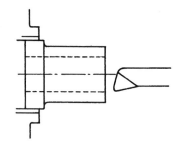

Figure 6.8 Boring.

A dial test indicator is set on the second centre, near the workpiece. As the workpiece is rotated the dial test indicator will show the degree of eccentricity of the part. The workpiece must now be adjusted until there is no movement of the dial test indicator indicating that the axis for turning is now in line with the machine axis. The dial test indicator should be placed so that it accurately shows the amount of adjustment of the chuck jaw.

The accuracy of the procedure depends on the accuracy of marking out and centre punching the part. If greater accuracy is required and if a threaded hole may be placed on the centre of the required diameter, then a procedure similar to button boring may be used (see Button Boring, page 47).

Boring operations

Boring is used to produce holes of an accurate size and true to the workpiece axis. The boring operation is preceded by a drilling operation to quickly remove unwanted material. The minimum amount of stock is left in the hole for removal by the boring tool. Because the boring tool is relatively small and usually poorly supported, the surface finish may be inferior.

If the required bore is a standard size, a better surface finish can be obtained by including a reaming operation. The process would then be:

1 drill out to leave 1 mm of material
2 bore out to leave about 0.5 mm of material (see Table 6.2)
3 ream the bore using a standard size reamer.

The purpose of the boring operation is to ensure that the hole is true to the axis of the work.

The reaming operation will then produce a bore of the correct size and of a superior surface finish. If the reaming operation is not completed at the same time as the boring operation, a floating reamer holder should be used.

It is important to include a facing operation at the same time as the boring operation so that a flat datum face is produced square to the bore. Boring tools may be solid or may use inserted cutters. Some boring bars allow adjustment of the cutters to ensure accurate bores.

Reaming after boring. If a reamer is used after a boring operation, it is essential that the tailstock is in line with the axis of the machine or 'bellmouthing' may take place. The extra load on the reamer may cause damage to the cutting edges.

If the reaming is not done at the same time as the boring, particular care is required to ensure that the bore is in line with the machine axis before reaming. A dial test indicator and a four-jaw chuck may be needed. Alternatively, a 'floating' reamer holder can be used. This holds the reamer firmly but allows some sideways movement so that the reamer can follow a bore that is not aligned with the machine axis. A floating reamer holder may also be required when the accurate alignment of the tailstock cannot be guaranteed.

Eccentric boring

The methods used for turning external diameters, are also suitable for producing eccentric bores. If a bore is to be set to an external datum, and if it must be accurately set, then button boring is the most suitable process.

Button boring. This procedure requires a setting 'button', which is a small cylinder with a known, accurate outside diameter and a central hole. The centre of the bore is marked out on the part and a small tapped hole produced. The setting button is accurately set in place using slip gauges and secured with a screw and washer. The workpiece can then be accurately set to the machine axis using a dial test indicator and 'clocking in' the setting button until there is no movement on the dial test indicator. The button is then removed and the boring operation performed.

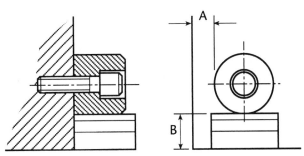

Set button to dimensions
'A' and 'B' with slip gauges

Figure 6.9 Button setting.

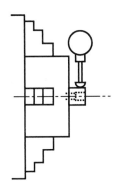

Figure 6.10 Setting button to machine axis.

Drilling and reaming

Holes may be drilled and reamed in the end of turned parts using the self-holding morse taper in the tailstock.

Straight shank tools are held in a Jacobs' chuck, larger tools with morse taper shanks are mounted directly into the tailstock.

Before drilling, it is essential that the tailstock is adjusted to be dead in line with the headstock. Use a dial test indicator mounted in the three-jaw chuck and rotated about the tailstock.

Drilling operations

Centre drilling (combined drill and countersink). These are used to create centres on the ends of shafts which are to be turned on centres, or can be used to provide an accurate guide for small drills.

After facing the shaft, select the best size of centre drill from Table 6.1 and mount in a Jacobs' chuck. Using a high-speed (min. 1200 rpm) and a slow feed drill about two-thirds of the length of the drill cone point. Typical faults include drilling too deep or a pilot length which is too short (see Fig. 6.12).

Table 6.1 Centre drilling (combined drill and countersink)

Work diameter	Drill number	Body diameter (in.)
Up to Ø6	1	1/8
Ø6–10	2	3/16
Ø10–12	3	1/4
Ø12–16	4	5/16
Ø16–22	5	7/16
Ø22–50	6	5/8
Ø50–75	7	3/4

Figure 6.11 Centre drill.

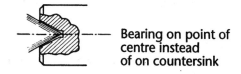

Bearing on point of centre instead of on countersink

Figure 6.12 Centre drill pilot too short.

Pilot drilling. This is a preliminary operation used when drilling larger holes (those over 12 mm). The chisel edge of the large drill must extrude the centre part of the work before cutting can start. This tends to force the drill off-centre and the hole may not be straight. The pilot drill is just larger than the chisel edge of the large drill and removes the centre portion of the work. The cutting force on the drill is then reduced and the hole is less likely to 'wander' off centre.

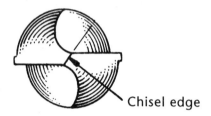

Figure 6.13 Pilot drilling.

Flat-bottom drilling. Pre-drill the hole with a standard type drill, taking account of the point angle follow with a flat-bottomed drill of the same size and remove the conical portion of the hole.

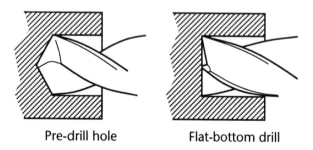

Pre-drill hole   Flat-bottom drill

Figure 6.14 Flat-bottom drilling.

Reaming

Reaming produces holes to H8 limits on the nominal size and gives a good surface finish. The small amount of stock removed means that cutting forces are low, but it is essential to avoid rubbing rather than cutting so a high feed rate is used.

As a rule of thumb, use half the speed used to drill the hole, but double the drilling feed rate. Do not reverse the rotation of the work or the reamer when retracting the tool. Soluble oil can be used to remove chips. or use an EP oil for a better finish. For cast iron use an air blast for chip removal (note safety precautions).

Drilling for reaming. The drilling operation leaves a small amount of stock for the reamer to remove. The amount of stock varies with the size of reamer, see Table 6.2. Too much stock causes heat, tool wear and a poor finish. Too little stock produces rubbing rather than cutting and rapidly blunts the reamer.

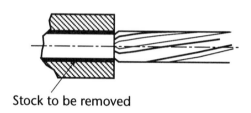

Stock to be removed

Figure 6.15

Table 6.2

Hole size	Stock on diameter (mm)
Up to ⌀10	0.25
⌀10–15	0.40
⌀15–20	0.50
⌀20–30	0.80
Over ⌀30	1.00

Taper turning

The compound slide

The compound slide is used to turn both large and small tapers. It has a limited accuracy, and hand feeding must be used, but it is a quick method of turning a wide range of tapers.

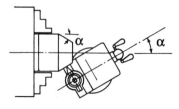

Figure 6.16 Compound slide.

The offset tailstock

This is used to turn shallow tapers with a high degree of accuracy. The machine feed may be used, producing a good surface finish. However, it takes longer to set up, and the taper cannot extend past the carrier. The offset must be calculated using half the included angle of the taper, and the length between centres. Alternatively, the taper is calculated as offset per unit length, and the tailstock adjusted using a mandrel and dial test indicator.

The taper turning attachment

The cross-slide is disconnected from the normal handwheel drive and attached instead to an adjustable angular slide at the back of the machine. Movement of the carriage displaces the cross-slide and the tool motion matches the angle of the slide.

While initial setting-up can be lengthy, no further adjustment is required and tapered parts can be produced quickly. Great accuracy can be achieved and the machine feed can be used giving good surface finish and rapid production. This method is most suitable for batch production work.

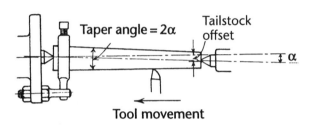

Figure 6.17 Offset tailstock.

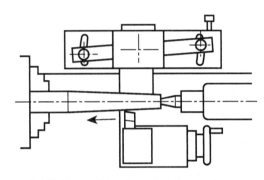

Figure 6.19 Taper turning attachment.

Forming tool

A rapid method of forming tapers, but the surface finish may be poor. Accurate tapers require the tool to be accurately produced and set, with consequent cost. General-purpose chamfers can be made quickly and cheaply using knife tools.

Inspection of tapers

Taper gauges

Taper ring gauges are used to inspect external or male tapers; taper plug gauges are used to inspect internal or female tapers.

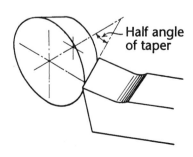

Figure 6.18 Form tool.

Figure 6.20 Taper ring and plug gauges.

Inspection procedure

Clean the workpiece and gauge thoroughly. Mark the male part with three equally spaced lines of chalk (turned parts) or engineer's blue (ground parts) along its length. **Do not use marking out blue.**

Assemble the tapers together and with light end pressure, rotate the male taper for a quarter of a turn, anticlockwise.

Check.

1 The diameter of the workpiece should lie between the **go/not go** steps of the gauge.
2 Try to rock the gauge, side to side. Any movement indicates an error on the taper. The end which rocks shows whether the taper angle is undersize or oversize.
3 Remove the male taper and observe the marked lines. A correct taper should show the lines rubbed all along their length. Rubbing at one end only shows an error on the taper.

It is important that mating tapers match each other in order to ensure correct location. In the case of self-holding tapers such as those in a lathe or drill spindle, errors on the taper could cause tools or holders to fall out from the machine spindle and cause injury. Always inspect tapers before assembly for nicks, burrs or damage. Do not use any tapered part which you think might be damaged or dangerous.

Tool height. An error in tool height setting will cause an oversize taper to be turned, even though all other settings are correct.

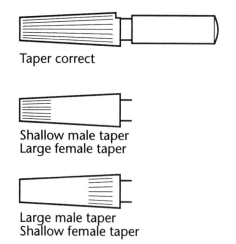

Taper correct

Shallow male taper
Large female taper

Large male taper
Shallow female taper

Figure 6.21 Checking a taper with 'blue'.

Standard taper systems

There are two basic categories of taper: quick release tapers and self-holding tapers.

Quick release tapers

These are used for location purposes only, usually in milling machine spindles. Retention may be by a drawbolt (see page 79). The taper angle is approximately 16.5°. The International and R80 systems are examples.

Self-holding tapers

These are used to locate the taper and to drive it by frictional grip. These tapers are shallow, less than 7°. They are used in drilling machine spindles, lathe headstocks and tailstocks. Metric, Browne and Sharpe & Morse are examples.

Screwcutting

Screwcutting involves both forming and generating processes. The thread profile is formed by the tool shape and the helix is generated by a combination of work rotation and tool movement.

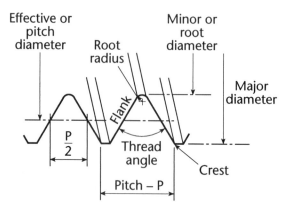

Figure 6.22 Thread terminology.

Standard thread forms

Triangular or vee threads

These are used for fastening or sealing. They are strong and relatively easy to produce. Production methods include milling, grinding, rolling, taps and dies, as well as screwcutting on the lathe. The basic geometry of the commonest forms is shown opposite. They use the pitch of the thread as the basis for working out the other thread dimensions.

For imperial threads

$$pitch = 1''/threads\ per\ inch$$

For metric threads the pitch is usually stated with the major diameter as follows:

$$M10 \times 1.5$$

which means the pitch is 1.5 mm.

If the pitch is not stated it indicates that the standard coarse pitch metric thread for the stated diameter is to be used. Consult a table of standard metric thread sizes.

It is recommended that the ISO metric screw thread should be regarded as the standard thread form.

Square threads

These are used to transmit motion or power (lathe leadscrew). They are stronger than vee threads and frictional forces are less, making transmission more efficient. The square form makes these threads awkward to cut and backlash, caused by wear, is difficult, to accommodate.

Trapezoidal and acme threads

These are used in preference to square threads for power and motion transmission. They are stronger and easier to cut than square threads and can be engaged and disengaged more easily. Frictional forces are higher than with square threads.

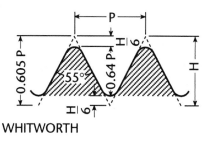

WHITWORTH

Figure 6.23

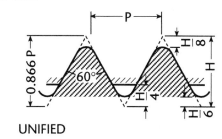

UNIFIED

Figure 6.24

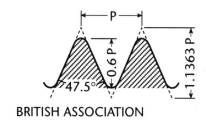

BRITISH ASSOCIATION

Figure 6.25

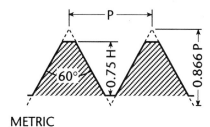

METRIC

Figure 6.26

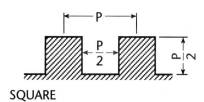

SQUARE

Figure 6.27

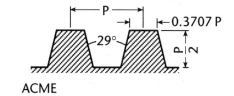

ACME

Figure 6.28

The ISO metric screw thread form

Metric thread tolerances

Both external and internal threads are toleranced in a similar way to limits and fits for shafts and holes. The tolerance zones are shown by letter and number combinations.

- External threads are shown as: 4h, 6g or 8g.
- Internal threads are shown as: 5H, 6H or 7H.

The class of fit is given by the appropriate combination:

Nut	Shaft	Class	Application
5H	4h	Close	Where precise accuracy is essential
6H	6g	Medium	For general purpose fits
7H	8g	Free	For quick and easy assembly for blind holes or where threads may be dirty.

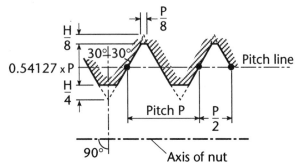

Figure 6.29 The ISO screw thread form.

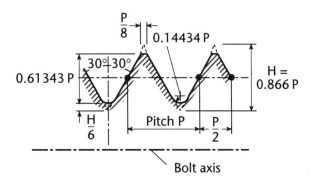

Figure 6.30

Screw cutting tools

These may be of HSS, carbide tip or carbide insert, with single or multiple points. When screwcutting on a manually controlled machine, the spindle speeds must be low so that the operator can withdraw the tool at the end of the cut. This makes it unsuitable for carbide tools. Carbides are best used on automatic or CNC machines where higher speeds can be used.

Like any form tool, the form ground on the tool must be adjusted to account for regrinding normal to the clearance angle and to accommodate any rake angle used.

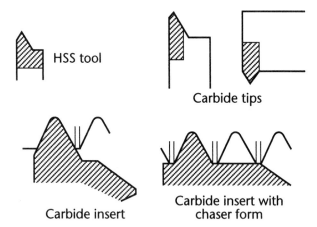

Figure 6.31 Screw cutting tools.

Infeed of the tool

The thread is cut with several passes of the tool. At each pass the tool is fed into the work a little more. This infeed may be square to the work axis (radial) or inclined to the axis (flank feed) using the compound slide.

Radial infeed. Radial infeed puts a heavier cutting load on the tool and creates more heat at the cutting edge. It is used with short chipping materials, with shallow threads, with square threads and with multi-start threads when pitch accuracy is essential.

Figure 6.32 Radial infeed

Flank infeed. Flank infeed reduces the load on the tool and gives better chip flow away from the workpiece. It is used with long chipping and difficult-to-cut materials. It is helpful with trapezoidal threads, which are commonly pre-machined with vee threads to reduce the cutting forces on the tool.

Flank infeed with the compound slide inclined at half the thread angle can produce rubbing on the trailing flank. Inclination at about 5° less than this angle reduces rubbing, improving surface finish at the cost of increased tool loading.

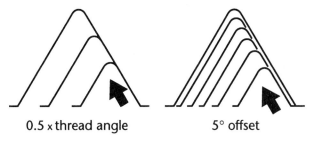

Figure 6.33 Flank infeed.

The compound slide movement must be calculated from the thread depth as follows:

Slide movement $S =$

$$\frac{\text{thread depth } D}{\text{cosin (half thread angle} - 5°)}$$

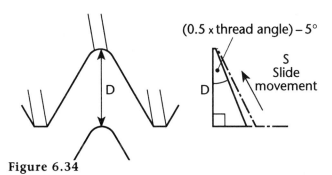

Figure 6.34

Tool angles

Tool side clearance. The tool must avoid rubbing the helical form of the newly cut thread. In practice side clearance is ground on the leading edge of the tool, of 3°–5° plus the lead or helix angle of the thread (see Fig. 6.35).

The lead angle is calculated from the lead of the thread and the circumference of the mean thread diameter.

Example calculations
Example 1
Tool side clearance angle. A thread of 5 mm lead has a mean diameter of 40 mm. Calculate the lead angle and tool side clearance angle.

$$\tan \text{lead angle} = \frac{\text{thread lead}}{\text{circumference of mean diameter}}$$

$$\tan \text{lead angle} = \frac{5}{\pi \times 40}$$

$$\tan \text{lead angle} = \frac{5}{125.66371}$$

$$\tan \text{lead angle} = 0.0397887$$

$$\text{lead angle} = 2.28°$$

tool side clearance angle
$$= 2.28° + \text{about } 4° = 6°$$

Tool side clearance angle = 6° (rounded off)

***Note*!** Normally the value of the thread pitch and the thread lead are the same. However, sometimes more than one thread may be cut on a shaft in order to give rapid movement of the nut. This is called a multi-start thread. For these threads

$$\text{lead} = \text{pitch} \times \text{starts}$$

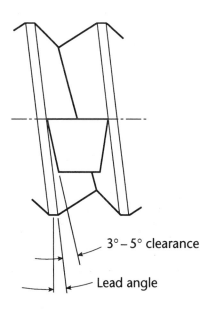

Figure 6.35 Tool side clearance.

Example 2

Thread lead. A two-start thread has a 5 mm pitch.

thread lead = pitch × starts
thread lead = 5 mm × 2 starts
thread lead = 10 mm

the lead value is used to calculate the tool side clearance.

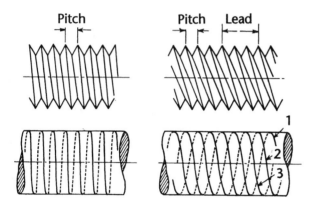

Figure 6.36 Thread pitch and thread lead (starts numbered 1, 2 and 3).

Tool side rake. If the tool has a tool side rake angle of 0°, the lead angle of the thread will cause the leading edge to cut with a working positive rake and the trailing edge to cut with a working negative rake.

If the tool is given a tool side rake angle equal to the lead angle of the thread, both edges of the tool will cut with a working rake angle of 0°. This will place the same cutting load on both flanks of the tool. The minor error on tool height can be ignored.

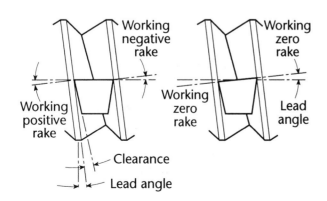

Figure 6.37 Tool side rake angles.

Tool and thread alignment

Cutting the thread with multiple passes requires the tool to engage the thread at the correct point for every pass. This can be done by keeping the carriage permanently engaged with the leadscrew, traversing back to the start point at the same feed as that used for screw cutting. This is obviously inefficient.

A quicker method is to engage the chaser dial with the lathe leadscrew and use the dial markings to ensure engagement at the correct point following tool disengagement the end of the cut. It is essential that a chaser dial gear appropriate to the required thread is used to engage the lathe leadscrew. This method allows rapid traverse back to the start point at the end of every cut, reducing production time.

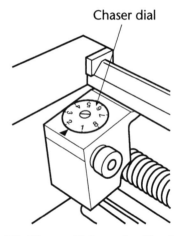

Figure 6.38 Chaser dial on the Harrison lathe.

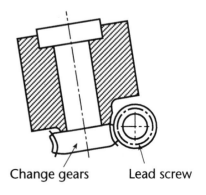

Figure 6.39 Chaser dial geared to the leadscrew.

Thread chasing

If it is desired to round off the crests of internal or external threads, a multi-point hand chasing tool is used in a light, trimming operation. The chaser is specific to a particular thread form and pitch. It is supported on a rest hold in the toolpost and lying alongside the cut thread. With the work rotating, it is pushed against the thread and allowed to feed itself along the screw thread. Several light cuts are taken checking the thread form each time against a gauge. The thread may be cut slightly large initially, to ensure a well-rounded form.

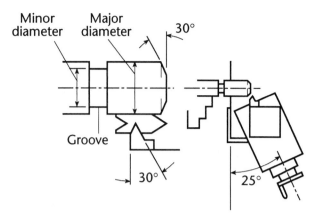

Figure 6.41

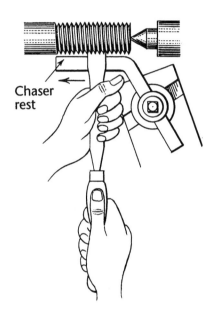

Figure 6.40 Hand chasing tool.

Single start screw threads – external

Right-hand vee external – metric

First calculate:

- thread major diameter
- thread minor diameter
- compound slide movement for 25° angle.

1 Secure the work on the machine. Turn the thread major diameter. Chamfer the shaft end, smaller than the thread minor diameter. Turn a groove at the thread end, less than the minor diameter.

2 Set the compound slide at 25°.

3 Set the tool on centre height, and square to the work. Clamp securely.

4 Touch tool on the major diameter. Set compound and cross-slide dials to zero.

5 Move tool to the end of the work, set depth of cut on the compound slide.

6 At a slow spindle speed and with ample coolant, engage the leadscrew, noting the chaser dial marking.

7 Disengage the leadscrew and stop the spindle when the tool enters the groove.

8 Withdraw the tool on the cross-slide, reposition at the start point, reset cross-slide dial to zero.

9 Set new depth of cut with compound slide.

10 Repeat the cutting procedure, engaging the leadscrew at the appropriate chaser dial marking each time. Stop just before the calculated depth.

11 Finish the thread with a die or chasing tool.

12 Check the thread with a ring nut.

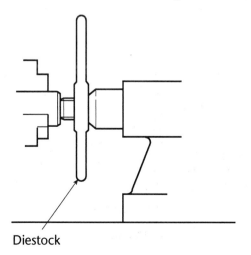

Diestock

Figure 6.42

Screw cutting using a die

A threading die may be used to cut large threads, from solid bar or to finish threads that have been screw cut on the lathe. Relatively large cutting forces are generated and particular care must be taken with long unsupported screwed shafts which tend to warp under the strain.

The die is assembled in the diestock and opened to give a roughing cut. The workpiece should be about 0.1 mm undersize and have a lead-in chamfer. One end of the diestock handle is set against the compound slide and the tailstock barrel is used to square the diestock body. The machine is isolated, and set in a high gear so that the chuck can be safely and easily turned by hand. As the die winds along the work, the tailstock barrel follows to maintain the die square to the workpiece. The chuck is reversed every half turn to break the chips.

Following the roughing cut, the die is closed to the finishing setting and a second cut taken. Mineral oil or tapping paste is used to help the operation.

Single start screw thread – internal

Right-hand vee internal – metric and left-hand vee internal – metric.
   First calculate:

- thread major diameter
- thread minor diameter
- compound slide movement for 25° angle.

1 Secure the work on the machine. Bore out to thread minor diameter. Chamfer the bore mouth larger than the thread major diameter. Turn a recess at the thread end larger than the major diameter.
2 Set the compound slide at 25°.
3 Set the tool on centre height, and square to the work. Clamp securely.
4 Move the tool into the recess and mark the saddle position on the bed slides, or mark the workpiece front face position on the boring tool. This indicates the position to disengage the leadscrew.

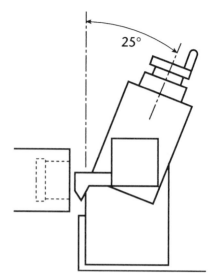

Figure 6.43 Slide setting.

5 Touch the tool on the bore and set compound and cross-slide dials to zero.
6 Move tool to the end of the work, set depth of cut on the compound slide.
7 At a slow spindle speed and with ample coolant engage the leadscrew, noting the chaser dial marking.
8 Disengage the leadscrew and stop the spindle when the tool reaches the recess, shown by the marked position.
9 Move the tool to the bore centre using the cross-slide. Reposition at the start point, reset cross-slide dial to zero.
10 Set new depth of cut with compound slide.
11 Repeat the cutting procedure, engaging the leadscrew at the appropriate chaser dial marking each time. Stop just before the calculated depth.
12 Finish the thread with a tap or an internal chasing tool.
13 Check the thread with an external screw gauge.

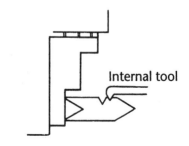

Figure 6.44 Tool setting.

For the left-hand thread, reverse the rotation of the leadscrew using the machine gearbox. Ensure that the lead angle clearance is on the correct face of the tool.

Using the procedure outlined above, cut the thread but start inside the recess and feed the tool towards the tailstock.

Alternatively, use the same procedure as for the right-hand thread, but use a left-hand screw cutting tool and reverse the direction of the rotation of the workpiece.

Ensure that the lead angle clearance is on the correct face of the tool. Check the orientation of the compound slide angle.

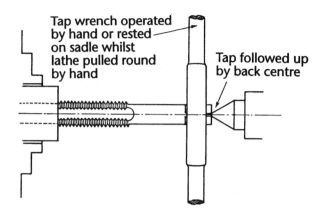

Figure 6.46 Tapping on the lathe.

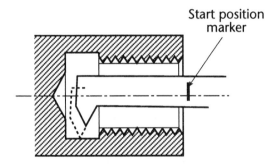

Figure 6.45 Tool marking.

Tapping internal screw threads

Standard internal vee threads may be more easily cut using hand taps. The appropriate tapping drill is selected and a hole of adequate depth drilled using the tailstock. The end of the hole should be chamfered to avoid a burr from the tapping operation. The taper tap is held in a tap wrench and is then set square to the workpiece using one of two methods.

If the tap has a centre hole, this is supported on the tailstock centre while the tap is engaged in the hole. The first two or three threads are then cut with this arrangement. The end of the wrench handle is set against the compound slide. The chuck is rotated by the left hand, the tailstock wound forward with the right hand.

If there is no centre hole, the tap is removed from the wrench and held in a Jacobs' drill chuck. This is mounted in the tailstock so that the tang just engages in the slot but the taper does not lock. The tap is engaged in the hole and two or three threads cut using the same procedure as above.

Once the tap has been started, it can be continued without support. The chuck should be reversed every half turn to break the chips. The thread is finished using the plug tap.

Acme and square threads

Right-hand acme external

First calculate:

- thread major diameter
- thread minor diameter
- compound slide movement for 14.5° angle
- thread lead angle

Ensure tool angles are correctly ground. Check tool form with a shadowgraph or toolmaker's microscope. If desired, the thread may

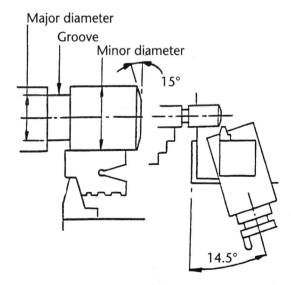

Figure 6.47

be roughed out with a vee thread tool to pitch × 0.5 deep, or with a square thread tool of width equal to 0.37 × pitch and depth equal to pitch × 0.5.

The thread is cut using the same procedure as that for the vee thread, except that the compound slide is set at 14.5°.

Check the thread with a single wire of diameter equal to thread pitch × 0.4872 and a micrometer. The reading over the wire should equal the thread major diameter.

Right-hand square external thread

First calculate:

- thread major diameter
- thread minor diameter
- thread thread depth – pitch × 0.5
- thread lead angle (lead = pitch × starts)

Ensure tool angles are correctly ground. Check tool form with a shadowgraph or toolmaker's microscope. If desired, the thread may be roughed out with a vee thread to a depth equal to pitch × 0.3. Reduce the major diameter by 0.25 mm.

The thread is cut using the same procedure as that for the vee thread, except that the compound slide is set square to the workpiece axis.

Check the thread with a single wire of diameter equal to the thread pitch × 0.5 and a micrometer. The reading over the wire should be equal to the thread major diameter.

Multi-start screw threads

Right-hand square external thread

The thread is cut using a similar procedure to that for the single start square thread, except that the compound slide is set dead parallel to the workpiece axis. Tool infeed is set using the cross-slide.

When the first thread has been cut, the tool is repositioned along the workpiece axis by a distance equal to one pitch, using the compound slide and ensuring that no backlash is introduced. The next thread is then cut following the same procedure.

Reposition the tool and cut further threads in a similar manner.

Instead of repositioning the tool for each thread form, the workpiece may be rotated to give the same effect. This is done by taking the work spindle gear out of mesh with the gear train, rotating the workpiece, then re-engaging the gear. The work spindle gear must have a number of teeth divisible by the number of thread starts, the gear teeth being used to indicate the correct degree of rotation.

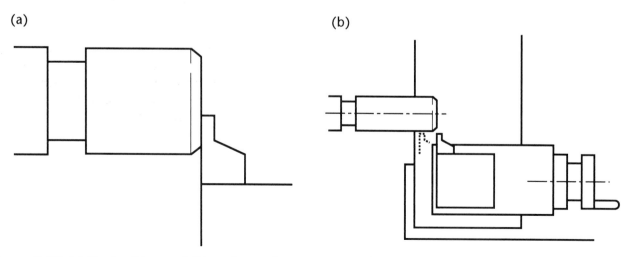

(a)   (b)

Figure 6.48 (a) Tool setting and (b) tool resetting – two start thread.

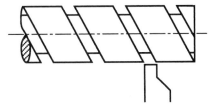

Figure 6.49a Cut first thread.

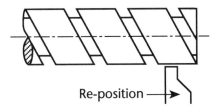

Re-position

Figure 6.49b

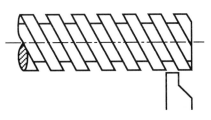

Figure 6.49c Cut second thread.

Safety when turning

1 Always use safety glasses and safety boots.
2 Tie up loose or long hair.
3 Do not wear loose items of clothing. Wear close fitting overalls and maintain them in good repair.
4 Do not wear rings or jewellery which could catch in the rotating workpiece.
5 Keep hands away from rotating parts.
6 **Always** use the guards provided, keep them properly adjusted and in good order. **Never** remove fixed guards unless authorised to do so.
7 Isolate the machine before changing workholding, checking dimensions or removing fixed guards around transmission items.
8 Make sure that workholding is correctly balanced.
9 Make sure that cutting tools, workholding and workpieces are securely attached to the machine.
10 **Never** handle swarf with bare hands, use a swarf hook.
11 Know how to stop the machine in an emergency. Check that the machine brakes are working properly.
12 Do not handle heavy workpieces or workholding unless you know the correct and safe procedures.
13 Clean the locating and/or seating faces of workholding equipment before mounting and after removing from the machine.
14 Ensure that chucks and chuck jaws are matched to each other, and to the machine where this is appropriate. Make sure that chucks are not run in excess of their designed speeds.
15 Ensure that the floor surrounding the machine is clean, sound and uncluttered.
16 Clean up oil and coolant spills immediately.
17 Store waste oils, coolant and swarf in the correct, designated containers.

7

Milling machines

Milling machines are generally identified by the type of construction and the orientation of the spindle. Machines may be classed as 'knee and column' or as 'bed' type, and the spindle may be horizontal or vertical. The main datum features of the machine must be correctly aligned to ensure accurate machining.

Additional features may include swivel tables, swivel heads and slotting attachments.

Milling operations use multi-tooth cutters to produce flat vertical and horizontal surfaces, open and closed slots and special forms. Milling operations are affected by the tool geometry, spindle speeds, table feed rates and cutter/workpiece movements.

Milling

Smaller milling machines are generally of the 'knee and column' type. The knee is wound up and down on an elevating screw, the column is fixed. Larger milling machines may be of the 'bed' type.

The cutting tools rotate and have many cutting edges (multi-tooth cutters). The work is fed past the cutter to produce the required machined form. On vertical spindle machines, the tool and spindle may be fed vertically for hole production.

Milling produces the following workpiece features:

- flat vertical surfaces
- flat horizontal surfaces
- closed or open slots
- grooves
- holes and bores
- cylindrical surfaces.

Types of milling machine

The vertical milling machine

Tools are mounted directly into the spindle, which is normally vertical. Straight shank tools are held in collets in a toolholder and retained by a drawbolt or cam-type clamp arrangement. Special toolholders are available to hold taper shank tools.

On some machines, the vertical spindle may be set at an angle for complex machining requirements. Attachments such as slotting heads provide further machining capability.

This is a very versatile machine and has largely superseded the older horizontal type machines.

Figure 7.1 Bridgeport milling machine.

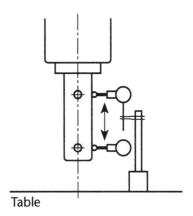

Table

Figure 7.2 Checking the spindle alignment.

Checking the spindle alignment

The spindle must be parallel to the knee slides and square to the table.

Check 1. Extend the quill (spindle housing), mount a dial test indicator on the table and traverse it up and down the front and side of the quill by winding the table up and down.

Check 2. Clamp a parallel plate on the table. Mount a dial test indicator off the spindle, set it in contact with the plate and rotate it. The head usually has some method of adjustment to correct any misalignment.

The horizontal milling machine

Tools may be mounted directly into the spindle, which is horizontal, or mounted on a spindle extension called an arbor. This is supported at its end to provide rigidity against the cutting forces.

This is a robust, production machine. It is fairly versatile and can machine a number of faces at one pass, however it cannot machine holes in the top surface of the workpiece.

Checking the arbor alignment. The arbor must run true between its housing and bearing support. Set a dial test indicator on an arbor spacer bush and rotate the arbor. Bent arbors must be discarded.

The universal milling machine

This is a horizontal milling machine on which the machine table can be swivelled for spiral/helical milling. Normally it is used with accessories, such as the dividing head, for the production of complex components or special operations. It is generally of lighter construction than the production-type horizontal machine.

Setting the swivel table parallel. The table must be set parallel to the knee slides for normal use. Clamp a parallel across the knee slides. Mount a dial test indicator off the table and traverse it across the parallel. Adjust the table until there is no dial test indicator movement.

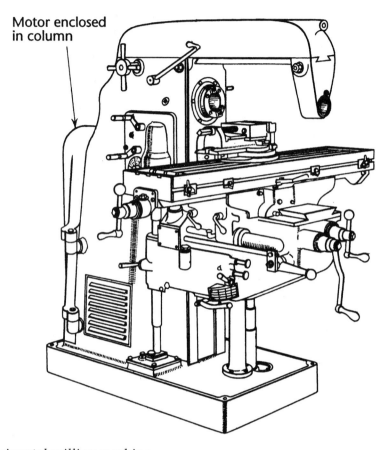

Figure 7.3 The horizontal milling machine.

Factors affecting milling

Milling operations are affected by such factors as cutting speed, feed rate, the direction of cut, the condition of the machine, as well as the machined form required.

The geometry of milling tools is more complex than that of lathe tools, and may require both axial and radial rake and clearance angles. The rates of speed and feed are recommended by the cutting tool manufacturer.

The direction of cut should oppose the workpiece feed for older machines or hardskinned materials, but may match the direction of feed if a backlash eliminator is available to give a better surface finish or to work with soft or thin workpieces. Selection of the appropriate type of cutting tool will produce the required machined form.

Milling cutter geometry

The geometry of a milling cutter is more complex than that of a turning tool. Milling tools

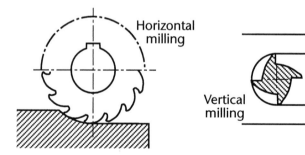

Figure 7.4 Cutter geometry.

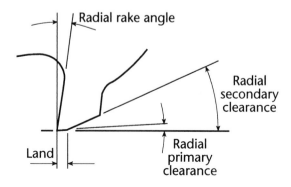

Figure 7.5

consist of a number of cutting edges (teeth) spaced radially around a cylinder. When cutting metal, this arrangement is similar to a lathe boring tool and each tooth must have a radial rake angle and radial primary and secondary clearance angles (see Fig. 7.5).

Because the cutter has more than one cutting edge, the tooth form is less robust than a similar single-point turning tool. Therefore the metal removal rate is reduced (smaller feed rate).

Cutters such as slot drills, end mills and face mills, also cut on the end of the tool. These end cutting teeth must also have tool angles for cutting and these may be referred to as the axial rake and axial clearance angles. Generally speaking the axial rake angle is determined by the helix angle of the tool.

When using HSS tools, the recommendations for rake angles follow those discussed in Chapter 9. As the workpiece material becomes harder or tougher then the rake angle decreases.

Short chipping or soft 'clogging' metals use a zero rake angle (see Fig. 2.9).

Carbide milling cutters

These are either solid carbide end mills or inserted tooth cutters. Inserted tooth cutters may be arranged to give various combinations of radial and axial rake. The following are the most widely used.

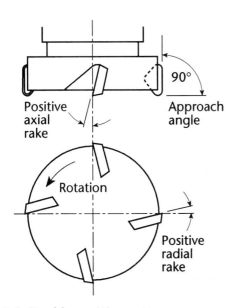

Figure 7.6 Double positive cutter.

Double positive. Positive radial and positive axial rake. Used on

- low-power machines
- light-duty operations
- low carbon steels
- heat-resisting steels
- stainless steels.

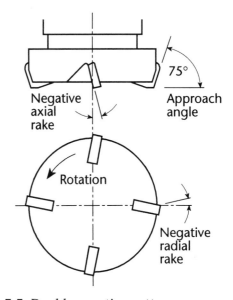

Figure 7.7 Double negative cutter.

Double negative. Negative radial and negative axial rake. Used on

- high-power machines
- high metal removal operation
- tough steels
- hard or abrasive materials

The double negative arrangement gives a very well-supported cutting edge, but requires plenty of power and a workpiece which is securely clamped.

In additional to radial and axial rake angles, the carbide insert shape may be presented to the workpiece in various ways. This is referred to as the approach angle, and will generally be 90°, 75° or 45°.

- The 90° approach angle produces square shoulders but 'hammers' the insert, and affects insert life and surface finish.
- The 75° approach angle gives good metal removal and a longer insert life.
- The 45° approach angle is best when dealing with the hard, abrasive surface of cast iron castings.

The inserts for milling cutters have flat edges (or facets) at their corners to give flatter, even surfaces. If turning inserts are used, the corner radius will produce a rougher surface finish. Where poor surface finish is due to the machine condition, a wiper blade may be used instead of one of the inserts. In this case, the inserts are used primarily for metal removal and the final surface finish is produced by the longer, convex cutting edge of the wiper.

Inserts are generally held in place by clamps, insertion into the body of the cutter giving substantial support to the cutting edge.

(a)

(b)

(c)

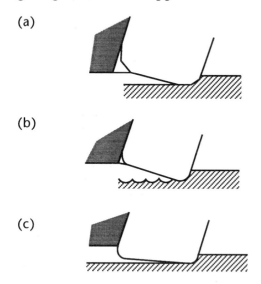

Figure 7.8 (a) Milling, (b) turning inserts and (c) wiper blade.

Flywheels

It is often beneficial to arrange for a flywheel to be incorporated into the toolholding arrangement, because of the fluctuating loads on the tool.

Cutting speed. This is the rate of material removal recommended by the tool manufacturer and is given as metres of chip per minute. It is developed experimentally and gives a good rate of cutting together with a reasonable tool life. Exceeding the recommended cutting speed substantially shortens the tool life.

The cutting speed varies with the workpiece and tool material. Cutting speed is related to the machine spindle speed in the following way

Spindle speed (rpm) (N)

$$= \frac{\text{Cutting speed (m/min) } (S) \times 1000}{\pi \times \text{diameter of cutter}}$$

or

$$N = \frac{S \times 1000}{\pi \times D}$$

Example calculations.

Example 1. A $\varnothing$20 mm HSS endmill is to cut a slot in mild steel (0.3% carbon steel) at a cutting speed of 30 m/min. Calculate the required spindle speed.

Spindle speed (rpm) (N)

$$= \frac{\text{Cutting speed (m/min) } (S) \times 1000}{n \times \text{diameter of cutter}}$$

$$\text{Spindle speed (rpm) } (N) = \frac{30 \text{ m/min} \times 1000}{\pi \times 20}$$

$$\text{Spindle speed (rpm) } (N) = \frac{30{,}000}{62.832}$$

$$\underline{\text{Spindle speed (rpm) } (N) = 477 \text{ rpm}}$$

Transposing the formula:

Cutting speed

$$= \frac{\text{Spindle speed} \times \pi \times \text{cutter diameter}}{1000}$$

or

$$S \text{ (m/min)} = \frac{N \text{ (rpm)} \times \pi \times D}{1000}$$

Example 2. The nearest available spindle speed is 420 rpm. What cutting speed would this give?

$$S \text{ (m/min)} = \frac{N \text{ (rpm)} \times \pi \times D}{1000}$$

$$S \text{ (m/min)} = \frac{420 \text{ (rpm)} \times \pi \times 20 \text{ mm}}{1000}$$

$$S \text{ (m/min)} = \frac{26389.378}{1000}$$

$$S = 26.4 \text{ m/min}$$

Milling feeds. Recommended feeds for milling cutters are given as feed per tooth. This varies with the workpiece material and the type of cutter used (due to the tooth form). Most machine feed mechanisms are set in terms of feed per minute and therefore a conversion must be made thus:

Feed/min = Feed per tooth ×
No. of teeth ×
Spindle speed (rpm)

or

$$F = f \times T \times N$$

Example 3. The endmill in Example 2 has four teeth and the feed per tooth is to be 0.1 mm. Calculate the feed per minute

Feed/min = Feed per tooth ×
No. of teeth ×
Spindle speed (rpm)

Feed/min = 0.1 mm × 4 × 420 rpm

Feed/min = 168 mm/min

Tool/workpiece movements

Up-cut milling

The direction of workpiece movement opposes the direction of relative tooth movement. Any backlash (free 'play' in the machine mechanism) is neutralised. The cutter tries to lift the workpiece, so rigid clamping is required. The load on the tooth is low at first and rubbing of the cutter edge may take place.

This method is normally used with older machines, or those without backlash eliminators. It is good practice with hard or abrasive skinned materials like cast iron castings, the cut being well advanced before it reaches the difficult area.

Down-cut milling

The direction of workpiece movement is in the same direction as relative tooth movement. Any backlash in the machine would allow the cutter to drag the workpiece beneath it, causing the tool to fracture. This method is only used with machines having no backlash or fitted with a backlash eliminator.

The cutting forces push the workpiece downwards to the machine table giving rigid support to the workpiece. This makes it particularly suitable for thin parts or those which are difficult to hold securely.

The power requirement is less (about 20%) so metal removal rates may be increased. The cutter is heavily loaded at the start of the cut, but the progression of the cut produces a better surface finish. This method of cutting is not suitable for parts which have hard or abrasive skins, such as cast iron castings, because this is the first area of tooth contact.

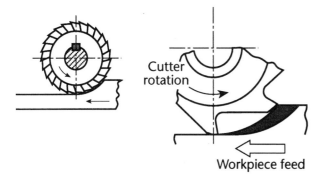

Figure 7.9 Up-cut milling.

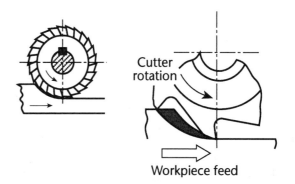

Figure 7.10 Down-cut milling.

Milling machine safety

1 Never operate any machine until there has been instruction and training in its safe use.
2 Regularly check that all machine controls are working properly.
3 Report defective or malfunctioning machinery at once. Where there is a safety hazard, isolate the machine and post warning notices.
4 Ensure that maintenance and remedial work is carried out properly and to schedule.
5 Never remove guards or protective equipment unless authorised to do so.
6 Always replace guards securely.
7 Know how to stop the machine in an emergency. Check that the machine brakes are working properly.
8 Ensure that the floor surrounding the machine is clean, sound and uncluttered.

8

Milling workholding

Workholding on the milling machine requires location, clamping and alignment of the workpiece. Workholding methods vary with the workpiece shape and the machining operation and include: direct clamping to the machine table, machine vices, vee blocks, angle plates, chucks and milling fixtures.

The dividing head and the rotary table enable more complex operations to be performed. The dividing head allows the machining of features with an angular relationship to each other. When coupled to the drive of the machine it allows spiral and helical forms to be generated. The rotary table allows the machining of circular slots and forms, and the production of features with an angular relationship to each other.

An essential element of workholding is accurate alignment of the workpiece, or workholding device, with the machine datum features and cutting tool.

Workpiece alignment

When milling, the cutter is normally held in a fixed position and the workpiece is traversed past it. The workpiece must not only be securely held, but must also be correctly and accurately aligned to the datum features of the machine. When considering workholding methods, attention must also be given to the procedures for ensuring accurate setting of the workpiece or workholding.

In most cases, a dial test indicator will be used to indicate the parallelism, squareness or concentricity of a datum feature on the workpiece or workholding. The datum will often be brought into alignment by tapping it with a soft-headed mallet, before final clamping takes place.

Workholding

Strap clamps

The simplest method of workholding is by

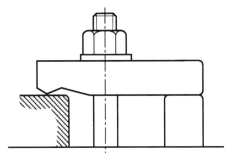

Figure 8.1 Strap clamp.

direct clamping to the machine table using strap clamps. A wide variety of workpiece shapes can be held using this method, including complex cast forms. An important feature on the workpiece is used as a datum for alignment purposes.

The clamp is made from a tough grade of steel and one end is relieved so that the clamping pressure is concentrated along a line of contact with the workpiece. The workpiece should be solid at the point of contact. Packing is placed at the rear of the clamp to give a large flat area of support.

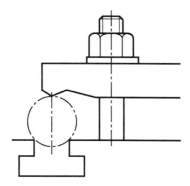

Figure 8.2 Location in the table tee slots.

The distribution of the clamping force varies with the position of the stud. Placing the stud as near to the workpiece as possible creates the best clamping arrangement.

This method is suitable for sawing operations and for milling slots or holes in flat workpieces. It may be necessary to move clamps when machining near the edge of the workpiece.

Strap clamps may be used to clamp small round bars located in the table tee slots. The diameter of workpiece should be no bigger than four times the slot width.

Cams and toggle arrangements may also be used to provide the clamping forces.

The machine vice

The plain vice. The most widely used method of workholding is the plain machine vice. It is particularly suitable for cold-rolled stock and machined parts such as square, rectangular, hexagonal and round bars.

The base and the seating face are made parallel and square to the fixed jaw. This allows the seating face and fixed (dead) jaw to be used as locating faces when setting-up for machining. When using these faces for location, the vice must be correctly set to the machine using a dial test indicator.

When a vice is to be used on a specific machine, tenons or keys may be attached to the vice base and adjusted so that they sit in the machine table tee-slots and automatically bring the vice square.

The fixed vice jaw will normally be set to oppose the cutting forces. If the fixed jaw is parallel to the cutting force, the workpiece is retained solely by the frictional grip of the vice jaws. This is not such a good set-up and should be avoided if possible.

Very long workpieces may require two vices to be set in line. The first vice is 'clocked in' as described above. A long parallel bar is extended from this vice and used as a setting bar to

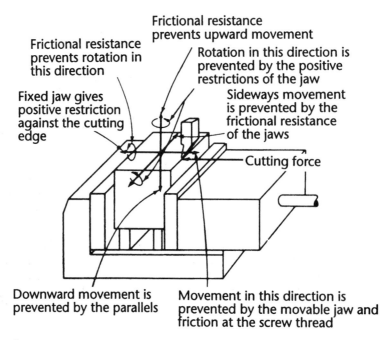

Figure 8.3 The machine vice.

position the second vice which is then clamped in position. The setting is checked using two matched parallels, one in each vice, set to abut each other. A dial test indicator is traversed along both parallels to show any deviation.

The vice may be set at an angle using a vernier protractor placed between the vice fixed jaw and the machine slides on the column.

In such cases, it may only be possible to use one tee nut and a strap clamp will also be required

The rotary or swivel vice. This is a plain vice set on a rotary base which enables it to be rotated to any angle about the vertical axis. A vernier scale enables accurate angles to be set.

The vice is set initially by rotating it to the '0' (zero) dial setting and then 'clocking in' as described above. This fixes the datum for angular rotation.

The universal vice. This is a rotary vice which may also be tilted about the horizontal axis. It is used for machining compound angles. It is set initially with both dials at '0' (zero).

Setting the workpiece in the vice

If the workpiece is geometrically true, it can be set on the seating face and against the fixed jaw and clamped firmly in place.

Clearances in the moving jaw mechanism can cause the live jaw to lift, tilting the workpiece. So, as the vice is tightened, the workpiece is driven down on the seating face using a mallet.

The mallet should not be allowed to bounce off the workpiece. A 'dead' sound indicates that there is a good seating between the vice and the workpiece. If the workpiece is lower than the vice jaw, it can be raised by placing a hardened steel parallel between the seating face and the workpiece. A paper 'feeler' placed between the workpiece and the parallel will indicate good contact if it is securely gripped.

If the workpiece is narrower than the parallel, a second parallel can be set between the live jaw and the workpiece.

Workpieces which are not quite geometrically true would not be held securely between parallel jaws and require a slightly different procedure.

The workpiece is cleaned and deburred and the largest, flattest face is set against the fixed jaw. A parallel may be used if required.

A round bar (about Ø20 mm) is placed between the live jaw and the workpiece and the vice tightened. The workpiece is tapped with a mallet so that it is firmly seated, paper feelers may be used to check this.

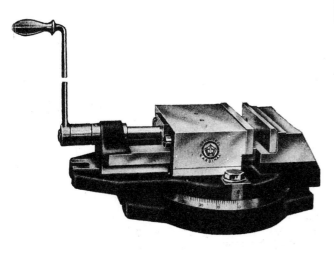

Figure 8.4 Rotary vice.

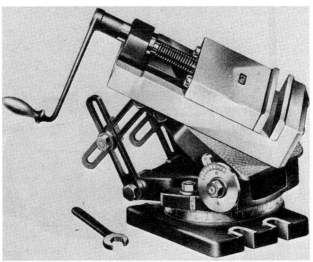

Figure 8.5 Universal vice.

Machining square and parallel

A workpiece say be machined square and parallel using the machine vice and a specific order of machining (see Fig. 8.6):

1 Set the workpiece in the vice using a round bar on the live jaw, with face X to the fixed jaw.
2 Lightly machine the top face, face Y.
3 Keep face X to the fixed jaw, seat on face Y.
4 Machine face Z to size.
5 Remove the round bar and clamp the workpiece on parallel faces Y and Z. Seat on face X using a parallel if necessary.
6 Machine face W to size.

The end faces can be machined with a side and face cutter or by setting the workpiece square in the vice using an engineer's square.

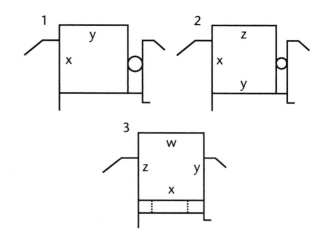

Figure 8.6 Milling sequence for squaring a block.

Vee blocks

Round workpieces are set into vee blocks to give clamping along three lines of contact. Vee blocks are supplied singly or in matched pairs. The datum faces on the vee block are used to set the workpiece parallel or square to the vice.

Vee blocks may be set directly on the machine table using strap clamps for fixing. The blocks are set in line using a datum edge formed by placing a parallel of the correct size in one of the tee slots. The tee slot is made to an accurate size for this purpose.

Vee blocks are supplied in a range of sizes to suit different workpiece diameters.

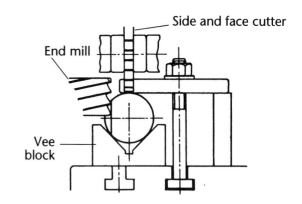

Figure 8.7 Vee block.

Angle plates

The fixed angle plate. Irregular-shaped work may be clamped to an angle plate for machining. Angle plates come in a range of sizes and are slotted for fixing purposes. The angle plate is set parallel or square using a parallel in a tee slot or using an engineer's square set against the table edge. If required, the plate may be set at an angle using a vernier protractor set against the slides on the column.

The workpiece may also be set at an angle using a vernier protractor or a sine bar and then clamped to the angle plate. Machining takes place into the angle plate.

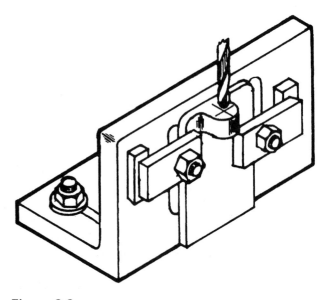

Figure 8.8

The swivel angle plate. This is a variation of the angle plate enabling large workpieces to be held securely at an angle. The plate is first set square or parallel with the swivel plate set at 90°.

Three- and four-jaw chucks

These are useful when working on the ends of round or hexagonal workpieces or for awkward shapes. The chuck may have an attached base plate with holes for fixing or may be held with strap clamps. The vertical spindle of the machine may be set central to round workpieces using a dial test indicator and attachment.

Milling fixture

This is a specialised workholding device usually designed to machine large numbers of specific components. The fixture holds the workpieces in a set position and provides for secure clamping.

A setting piece is provided so that cutters can be accurately set to the fixture using feeler gauges of a specific size.

The dividing head

The dividing head allows machining of features which have an angular relationship to each other, e.g. holes equally spaced on a pitch circle diameter (PCD), or a hexagonal form milled on the end of a shaft.

It can also be used to machine circular forms and is used in conjunction with the milling machine drive mechanism to generate helical/spiral forms. It is held on the machine table with tee bolts. If the dividing head is to be used vertically, it is 'clocked in' centrally to a round bar using a dial test indicator and attachment.

If it is to be used horizontally, it is set using a parallel mandrel in the taper bore in the head. This is first checked for true running by rotating it against a dial test indicator The head is then set parallel by traversing the dial test indicator along the mandrel both on its top and on its side.

Long workpieces will require support from the tailstock.

The dividing head may be set at an angle using the integral vernier protractor. The tailstock may be set at an angle using a vernier protractor seated against the machine table.

The plain dividing head

This allows workpieces to be rotated through fractions of a complete turn using manual methods. The operations are described as either direct or simple indexing.

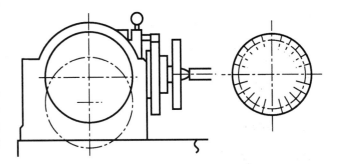

Figure 8.9

Direct indexing. The main spindle of the head carries a disc which has an equally spaced circle of holes or serrations. A plunger engages with these holes (or serrations). When the plunger is disengaged, the spindle and workpiece can be freely rotated. The hole spacing allows the operator to rotate the workpiece through a specific angle or a specific fraction of a turn.

Figure 8.9 shows a plain dividing head with a disc having 24 holes. Other hole patterns commonly available are 20, 42 and 60 holes.

The hole spacing can be used in two main ways:

- to space holes on a PCD
- to index through a specific angle.

Examples of direct indexing

Example 1. Hole spacings on a PCD.

Holes on PCD	Plunger engages every X hole on dividing disc
24	First
12	Second
8	Third
6	Fourth
4	Sixth
3	Eighth
2	Twelfth

Example 2. Indexing through an angle. This is based on the fact that the smallest rotation is 1/24 of a turn and 1/24 turn equals

$$\frac{360°}{24} = 15°$$

Index angle (degrees)	Plunger engages every X hole on dividing disc
15	First
30	Second
45	Third
60	Fourth
75	Fifth
90	Sixth
105	Seventh

Simple indexing. The main head spindle and workpiece are connected to a worm and wheel mechanism. On the end of the wormshaft is a crank, allowing manual rotation of the worm.

The usual worm/wheel ratio is 40:1 so that the crank must be turned through 40 turns in order to turn the workpiece through one complete turn, or 360°. Fewer turns of the crank give a smaller angle of index.

Turns of the crank	Angular rotation of the workpiece
40	360°
30	270°
20	180°
10	90°
1	9°

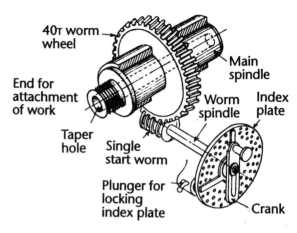

Figure 8.10 Worm and wheel mechanism.

$$\text{Angular rotation} = \frac{360° \times \text{turns of crank}}{40°}$$

$$\text{Turns of crank} = \frac{\text{required angle}}{9°}$$

Alternatively the rotation could be expressed as a fraction of a turn thus:

Turns of the crank	Fractional rotation of the workpiece
40	1 turn
30	3/4 turn
20	1/2 turn
10	1/4 turn
1	1/40 turn

$$\text{fractional rotation} = \frac{\text{turns of the crank}}{40}$$

Sometimes the required angle or fraction of a turn cannot be achieved with whole turns of

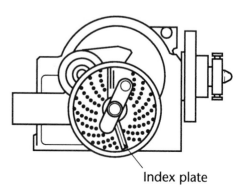

Figure 8.11 The dividing head index plate.

the crank. In this case it may be necessary to rotate the crank through only part of a turn. The indexing plate next to the crank is used to give part turns of the crank, using the same basic principle as direct indexing.

The indexing plate has a number of equally spaced circles of holes. The number of holes in each circle is marked on the plate.

A typical selection might be

15, 16, 17, 18, 19, 21, 29, 33, 39, 43, 49, 50, 54

The crank plunger pin is adjusted to engage with a hole circle which can give the required fraction of a turn. For example: if the crank must be moved by a 1/2 turn, the selected hole circle must be divisible by 2. Select the smallest hole circle – 16. Disengage the crank plunger pin, move eight holes around the 16-hole circle and then re-engage the plunger pin.

Examples of simple indexing

Example 1. Angular rotation. To index 12°

$$\text{Turns of crank} = \frac{12°}{9°} = 1\frac{3}{9} = 1\frac{1}{3} \text{ turns}$$

Select a hole circle divisible by 3.

Select a 15-hole circle.

Indexing move = one complete turn and five holes on a 15-hole circle.

Example 2. Hole spacing. To have six equally spaced holes

$$\text{Turns of the crank} = \frac{40 \times 1}{6} = \frac{40}{6} = 6\frac{4}{6}$$

$$= 6\frac{2}{3}$$

Select a hole circle divisible by three.

Select a 15-hole circle.

Indexing move = six complete turns and ten holes on a 15-hole circle.

The universal dividing head

The universal head takes simple indexing a stage further and allows rotation of the workpiece to a higher level of precision or for a wider range of indexes. This is done by unclamping the index plate and connecting it by gearing to the main spindle. As the crank rotates the spindle and the workpiece, the index plate also rotates a small amount.

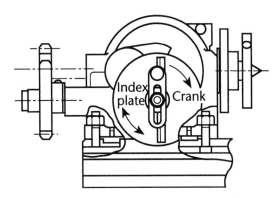

Figure 8.12 The universal head.

This differential movement allows the angle of rotation to be either increased or decreased a little. There are various methods of using this facility, called compound, differential and multiple indexing.

The main spindle may also be connected by gearing to the machine leadscrew so that as the machine table feeds forward, the workpiece rotates. This allows machining of helical and spiral forms.

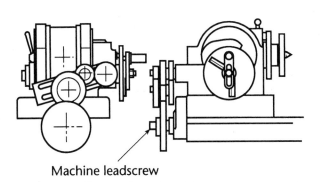

Machine leadscrew

Figure 8.13 The universal dividing head.

Example of differential indexing. It is required to drill 61 holes around the outside of a bar. Calculate the indexing move and gearing required.

Hole circles available:

15, 16, 17, 18, 19, 21, 29, 33, 39, 43, 49, 50, 54.

Gears available:

4T, 32T, 48T, 54T, 60T, 72T, 84T, 96T, 100T.

Nearest simple index move would be for 60 holes, giving an indexing move of

$$\frac{40}{60} = \frac{2}{3} \text{ turns}$$

Simple index move = ten holes on a 15-hole circle.

But this gives 60 holes instead of 61 holes. The gearing must reduce each index slightly so that an extra drilled hole can be introduced

Gear ratio required = simple index move × No. of holes extra or fewer

$$= \frac{2}{3} \times 1 \text{ hole}$$

$$= 2 : 3$$

Select suitable gears to provide this ratio. Gears require 32T:48T.

The rotary table

This is similar in principle to the rotary vice, except that a flat surface with tee slots is provided for clamping purposes. The angle of rotation may be finely set using an integral vernier protractor.

It is used for milling angular faces and circular arc forms and can be used with various workpiece shapes.

A central bore is provided so that a spigot can be used, with a dial test indicator, to bring the table central to the machine spindle for initial setting. The table can be clamped in place or can be used to rotate the workpiece beneath the cutter using a worm and wheel drive.

The table rotates about the vertical or horizontal axis. It has tee slots in the table for workholding and is usually supplied with an accurate centre hole for alignment purposes.

The table is driven through a worm and wheel arrangement. It may be possible to disengage the worm drive for rapid rotation of the table.

Rotation is controlled by a handwheel and most rotary tables have a vernier scale for setting the table within 5" of angular rotation. Some rotary tables use an indexing plate, similar to the dividing head.

Various worm wheel ratios are used including

40:1, 72:1, 80:1, 90:1 and 120:1.

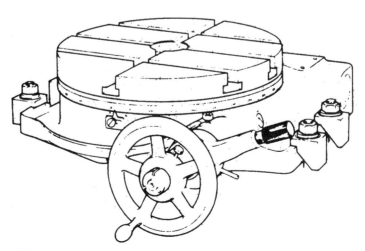

Figure 8.14 The rotary table.

Figure 8.15 Rotary table vernier.

The calculation of the number of turns required is similar to that used with the dividing head

$$\text{No. of turns of the crank} = \frac{\text{angle required}}{(360°/\text{No. of teeth in worm wheel})}$$

Workholding safety

1 Check the condition of workholding before and after use. Remedy or report faults.
2 When lifting or transporting workholding, ensure that the correct procedures are known and used.
3 Use lifting and handling equipment wherever possible.
4 When a two-person lift is necessary, both people should be of the same height and they should agree which person will direct the operation.
5 Ensure that traffic routes are clear and sound.
6 Prepare a safe, accessible and secure resting place for workholding equipment before handling and transport operations begin.
7 Check that workholding is correctly and securely attached to the machine before it is used.
8 Remove workholding when it will not be used for some time, to avoid corrosion taking place.
9 Clean and lightly oil workholding after use and before storage.
10 Ensure that the storage method is safe.

9

Milling operations

Cutting tools for horizontal milling machines are distinguished by the central bore necessary for mounting on the machine arbor – normal or stub.

Long arbors are prone to bending and must be well supported when mounting, cutting, dismounting and storing. Knee braces may be required for heavy cutting operations.

Standard cutters produce horizontal and vertical machined faces and may be mounted in various combinations to produce complex forms at a single pass.

Special cutters are available to produce a wide range of complex forms.

Cutting tools for vertical milling machines have a plain or threaded shank so that they can be held in a split collet.

Collet chucks may be operated by drawbar, nut and spanner or by using quick-acting cam forms.

Standard cutters produce horizontal and vertical machined faces and bores. Angular faces can be produced with standard tools if the head can be inclined at an angle.

Special cutters are available for special forms such as tee-slots and dovetail slots.

Boring heads are used to machine very accurate bores.

Horizontal milling – cutting tools

Arbor mounted cutters

These cutters contain a hole for mounting purposes and a keyway for driving purposes.

Plain cylindrical cutters. Plain cylindrical cutters, or slab mills, produce wide, flat surfaces at one pass. Helical teeth give a much smoother cutting action than straight teeth and help with chip removal. An end thrust is introduced by the helix and, where possible, it is arranged for this to be directed towards the column during cutting.

Light-duty cutters have many teeth but a slow helix (20°). Few teeth are in contact with the work, so cutting forces are low. This makes them suitable for older, low-powered machines.

Heavy-duty cutters have fewer teeth but a larger helix angle (35°). More teeth are engaged during cutting, giving a better cutting action and larger rates of metal removal.

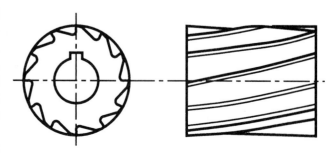

Figure 9.1 Light-duty cutter.

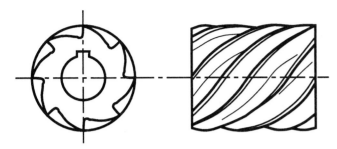

Figure 9.2 Heavy-duty cutter.

Helical mills. These are similar to plain mills but there are generally only three or four teeth and the helix angle is 45° or greater. They are particularly suitable for heavy cutting operations or the machining of rough surfaces. However, with a very large helix angle, the smoothness of cut also makes them suitable for light metal removal on thin sections which might otherwise deflect under the tool.

Figure 9.3 Helical mill.

Side and faces cutters. These cut on the outside diameter and the sides of the teeth. They produce both horizontal and vertical surfaces, and slots.

Interlocking cutters are used for faces which are wider than a single cutter width. The teeth of the cutters overlap slightly.

Staggered tooth cutters cut on alternate sides of the cutter. They are used for deep slots and give a good surface finish.

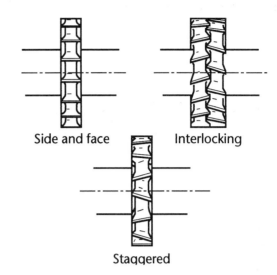

Side and face   Interlocking

Staggered

Figure 9.4 Side and face cutters.

Angular cutters. These produce angled faces. Applications include workpiece chamfers and cutting the flutes or tooth forms on other cutting tools.

Double angle cutters. These are used to mill angular grooves, such as vee forms in workpieces. They are also used to form flutes (tooth forms) in helical flute cutting tools, where the ordinary angular cutters would interfere with the helical flute as it was cut.

Saws. These cut on the outside diameter only, side clearance being ground on the tool. They are used to machine narrow slots and to cut material into pieces. They are provided in narrow widths of 0.5 mm upwards. Staggered tooth cutters are wider but give a better cutting action and surface finish. Screw slotting cutters are very thin and as the name suggests, produce narrow slots in screw heads.

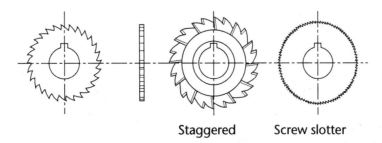

Staggered   Screw slotter

Figure 9.5 Saws.

Form relieved cutters. These are a special type of cutter, used to form profiles on the workpiece. They are sharpened by regrinding the front face of each cutting edge, not the outside diameter, so maintaining the correct cutter profile.

Form relieved cutters include:

- convex cutter
- concave cutter
- single radius cutter
- double radius cutter
- fluting cutter
- gear cutter
- gear hob.

Stub arbor mounted cutters

These provide a means of machining wide, flat vertical surfaces on the component. The cutter locates on a short spindle extension and is retained by a screw or nut. Drive from the spindle is by means of keys or 'dogs'. Cutting takes place on the end teeth and the outside diameter.

A distinction is made between smaller cutters called shell end mills (∅30–150 mm) and the larger face mills (over ∅150 mm).

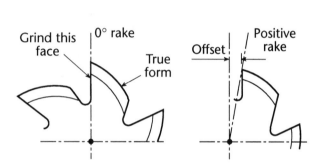

Figure 9.6 Grinding form relieved cutters.

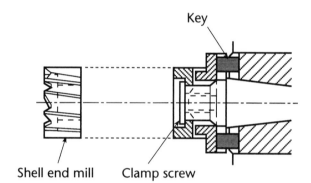

Figure 9.8 Stub arbor mounted cutters.

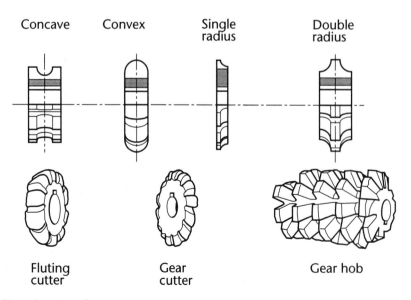

Figure 9.7 Form relieved cutter forms.

Horizontal milling – toolholding

Machine arbors

The arbor is a long extension to the machine spindle. It carries a range of cutting tools and is subject to bending forces during cutting. If the arbor becomes permanently bent it should no longer be used. Care must be taken to support the arbor adequately during cutting, and when not in use it should be hung vertically in a storage rack. Periodically the arbor should be checked in the machine for distortion using a dial test indicator.

The arbor locates in the machine spindle by means of a quick-release taper of about 16° (inclusive). It is retained by a long bolt called a drawbolt and is supported at its extreme end by an arbor support and a support bearing.

The arbor, the support bearing and the spacer bushes must be cleaned and inspected for burrs and nicks before the arbor is mounted on the machine.

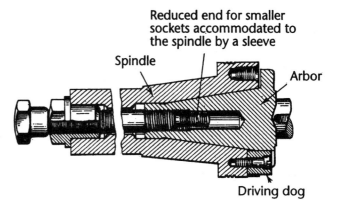

Figure 9.10 Arbor location and retention.

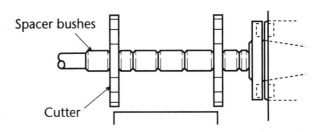

Figure 9.11 Cutter spacing.

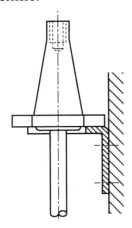

Figure 9.9 Arbor storage.

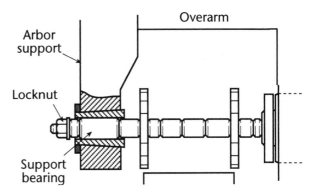

Figure 9.12 The arbor support.

Mounting the arbor. The arbor is inserted into the spindle taper and securely located on the drive keys. The drawbolt is then tightened, the arbor being supported by hand.

The cutter is positioned on the arbor using spacer bushes. Generally the cutter is set as close to the column as possible, while still being capable of completing the required machining operation. A drive key is inserted between the cutter and the arbor.

Additional spacer bushes are then added to position the support bearing. This is generally placed as close to the cutter as possible, but clear of the workpiece and workholding equipment. The arbor support is then positioned over the support bearing, and locked in place on the machine overarm. The overarm itself may be repositioned if required to accommodate varying arrangements of cutters.

Additional spacer bushes are then added to position the arbor locknut over the thread. This is then tightened firmly.

Note! The arbor locknut must **never** be tightened or loosened unless the arbor support is in position over the support bearing. Otherwise the arbor will be twisted and permanently distorted.

Knee braces. These provide additional support against cutter, arbor and overarm deflection when several cutters are being used together (gang milling).

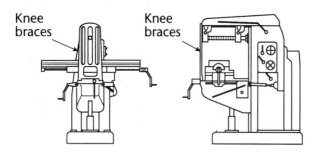

Figure 9.13 Milling machine knee braces.

Stub arbors

This is a shortened form of the normal arbor and is used for direct mounting of shell end mills and face mills. The taper must be clean and free from burrs and nicks, and the drive keys securely engaged before the drawbolt is fully tightened.

The cutter must be truly located on the arbor and fully engaged with the arbor drive keys or drive slots, before the clamping nut or screw is fully tightened (see Figs 9.8 and 9.14).

Horizontal milling – machining operations

The horizontal milling machine can be used to produce vertical, horizontal and angular faces, including the production of several faces at a single pass. Slab mills and helical mills produce wide, flat horizontal surfaces. Angular faces can be produced by setting over the workpiece at the required angle before machining, or by using special angular form cutters. Shell end mills and face mills produce wide vertical faces and shallow vertical surfaces (see Fig. 9.16).

Side and face cutters produce both vertical and horizontal surfaces. They can be mounted as shown to produce parallel faces, this is called straddle milling. The cutters must be matched for diameter to ensure the same depth of cut. The width between the cutters is set using spacers and shims and checked using

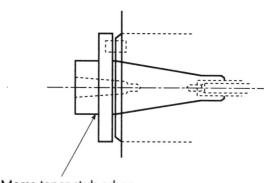

Morse taper stub arbor

Figure 9.14 Stub arbor.

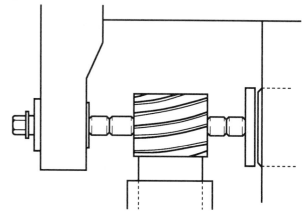

Figure 9.15 Slab milling horizontal surfaces.

slip gauges. The side of the cutters may be checked with a dial test indicator to ensure true running (see Fig. 9.17).

Side and face cutters can be used to cut vee forms on a diameter by offsetting the cutter to the axis of the workpiece.

A combination of side and face cutters and slab mills can be used to produce complex forms at one pass. This is called gang milling.

Figure 9.16 Milling wide vertical surfaces.

Some or all of the cutters will have to be specially made to the required dimensions.

The cutting forces are very high and knee braces may be required.

The slab mills are arranged so that axial forces, due to the cutter helix, either cancel each other or are directed toward the machine column (see Fig. 9.18).

Special form cutters reproduce the cutter form in the workpiece. The main task is to ensure that the cutter and workpiece are accurately set to each other. A ground parallel may be clamped next to the cutter on the arbor and a feeler gauge used to set this to the workpiece. The cutter will be central to the workpiece when it is traversed a distance equal to the feeler, the parallel and half the width of the cutter.

Figure 9.18 Gang milling.

Vertical milling – toolholding

Collect chucks. These are generally used with plain or threaded shank tools. A range of standard sized split collets are used to grip the tool shank, the collets being 'pull' type and operated by a drawbar, through the machine spindle.

The drive from the spindle to the collet is by a small shear pin, but the drive from the collet to the tool only uses friction. This limits the maximum size of cutter that can be used (⌀50 mm). Tool changing can be time consuming.

Figure 9.17 Straddle milling.

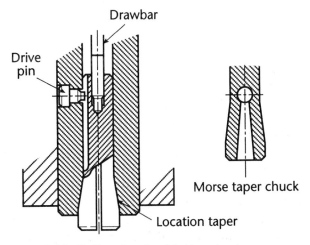

Figure 9.19 Collet chucks, frictional grip.

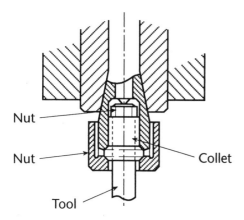

Figure 9.20 Collet chuck, positive grip.

Some toolholders are available which lock the tool in place using a special nut on the tools threaded shank. The tool is gripped between the collet and an internal centre, and cannot screw itself out of the collet. Tool changing is quite rapid.

The toolholder is held in the machine spindle by a drawbolt and driven by fixed keys ('dogs'). Changing the toolholder can be time consuming.

Quick-change toolholders are available which operate a 'push'-type collet using a special nut. The toolholder is held in the machine spindle by a quick acting cam-type clamp arrangement and driven by 'dogs'. Changing tools and toolholders is rapid.

Stub arbors. Larger vertical milling machines may accept stub arbor-type toolholders, allowing the use of large shell endmills and face mills.

Vertical milling – tools and operations

The vertical milling machine is used to produce vertical, horizontal and angular faces. Special tools are available to produce complex forms at a single pass.

End mills. These cut on the outside diameter and the end face and may have three or four flutes. Endmills produce horizontal and vertical surfaces, open-ended slots, profiles and pockets. They cannot be 'plunged' into the workpiece like a slot drill. The outside dia-

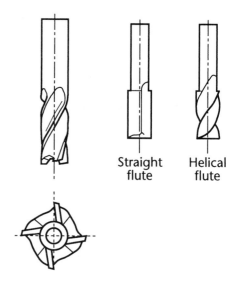

Figure 9.21 Four-flute endmills.

meter is not made to a close tolerance, so accurate slots need more than one pass and accurate machine setting. They are reground on the diameter and end faces.

Slot drills. These cut on the outside diameter and the end face. They are made to close tolerances and can be 'plunged' into the workpiece like a drill. Slot drills produce horizontal and vertical faces, open and closed slots, profiles and pockets, and sunken or recessed forms. Their accurate width enables keyways to be cut at a single pass.

Ball-nosed slot drills have a spherical end and are used for milling complex forms in dies and moulds, and radiusing internal corners. Slot drills are reground on the end faces.

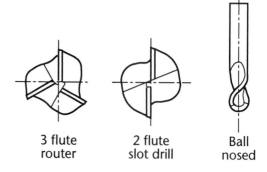

3 flute router   2 flute slot drill   Ball nosed

Figure 9.22 Slot drills.

Router cutters. These are modified slot drills, having three flutes. They can be used for the standard slot drill operations. Metal removal rates are higher, but the surface finish may be poorer.

Endmills and slot drills can produce angular faces when the workpiece is set over at the required angle.

It is common practice to use up-cut milling for roughing operations followed by down-cut (climb) milling for finishing. This gives a better surface finish but can only be done when there is no backlash (play) in the machine mechanism.

Milling slots in solid metal involves both up-cut and down-cut milling giving a balanced machining arrangement.

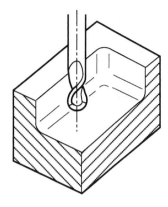

Figure 9.25 Ball-nosed slot drill.

Woodruff cutters. These are made to set diameters and widths and are used to cut circular slots in shafts for woodruff keys. They only cut on the outside diameter and should not be confused with tee slot cutters.

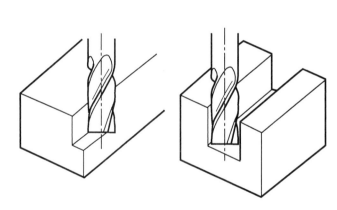

Figure 9.23 Slot drill and endmill operations.

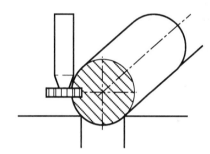

Figure 9.26 Woodruff cutter.

Tee slot cutters. Tee slot cutters are made in a range of standard sizes corresponding to standard tee slots. They produce the slot at one pass, cutting on the top and bottom faces and the diameter. An endmill or slot drill must first be used to cut a slot large enough to allow the 'neck' of the tee slot cutter to pass freely through the workpiece. High spindle speeds and low feed rates reduce the cutting forces on the tool.

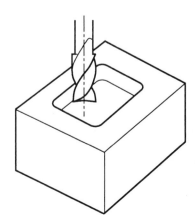

Figure 9.24 Slot drill operation.

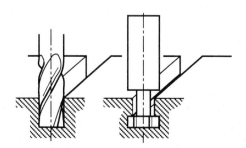

Figure 9.27 Tee-slot cutter.

Flycutters. These have been used in the past to cut wide, flat surfaces, a single point tool being arranged in a suitable holder. The development of large face mills, and the danger of tools being flung from the flycutter, have reduced the use of these tools.

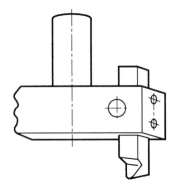

Figure 9.28 Flycutter.

Dovetail cutters. These are used to produce standard internal or external dovetail forms for slideways in machines, small tools and fixtures. The dovetail angle depends on the cutter and is available over a restricted range usually 45°, 55° and 60°. Rollers are used to measure the dovetail form.

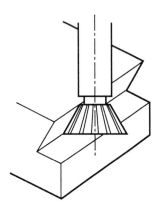

Figure 9.29 Dovetail cutter.

Shell endmills and face mills. Shell endmills and face mills produce wide, flat horizontal surfaces. The side of the cutter can be used to produce vertical or angled faces. They are available with a variety of rake angles and approach angles. These tools remove large areas of metal rapidly, but require more power than HSS cutters and the workpiece must be rigidly held. See also Chapter 7.

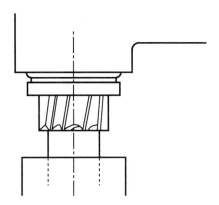

Figure 9.30 Facemill.

Vertical milling machines can also be used to perform the normal range of drilling, reaming and boring operations.

The offset boring head

A single-point boring tool is set in the head and used to open out a previously drilled or cored hole. The tool can be adjusted inwards or outwards in small increments of 0.025 mm (0.001″) enabling accurate bores to be produced. The tool is offset using a fine pitch screw arrangement as in a micrometer.

The operation is similar to boring in a lathe, and requires low rates of metal removal and the use of 'spring' cuts to ensure dimensional accuracy.

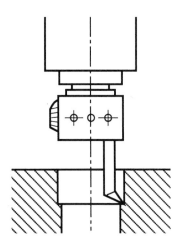

Figure 9.31 Offset boring head.

Safety precautions when milling

1 Use safety glasses.
2 Tie up loose or long hair.
3 Do not wear loose items of clothing. Wear close-fitting overalls and maintain them in good repair.
4 Do not wear rings or jewellery which could catch in the rotating cutter.
5 Keep hands away from rotating parts.
6 **Always** use cutter guards and keep them properly adjusted.
7 Isolate the machine before changing the cutter, checking dimensions or adjusting drive belts.
8 Wrap cutters in cloth when moving them to and from the machine.
9 **Never** handle swarf with bare hands, use a swarf hook.
10 Know how to stop the machine in an emergency. Check that the machine brakes are working properly.
11 Lower the machine table to elbow height when mounting or removing workholding equipment.
12 Clean the table and the seating faces of workholding equipment before mounting and after removing workholding equipment.
13 Ensure that the floor surrounding the machine is clean, sound and uncluttered.
14 Clean up oil and coolant spills immediately.
15 Store waste oils, coolant and swarf in the correct, designated receptacles.

10
Grinding

Grinding is a machining process, using an abrasive grain as the cutting tool. It is particularly suitable for hardened workpieces or where smooth, accurate forms or faces are required.

The double-ended off-hand grinding machine has rough and smooth wheels for manual grinding of small tools and workpieces.

The surface grinding machine produces flat or formed external surfaces. The spindle may be vertical or horizontal and the workpiece may be reciprocating or rotating. Production rates range from one-off to high volume. Workholding is most often by magnetic chuck, but may also be by direct clamping, machine vice or grinding fixture.

The cylindrical grinding machine produces internal and external diameters and tapers, and flat shoulders and end faces. Formed wheels can produce formed workpieces. Workholding may be by three- or four-jaw chuck, centres, collets, mandrels, magnetic chuck or grinding fixture.

Grinding wheels must be balanced and dressed by qualified and authorised personnel. Particular attention must be paid to safety, especially eye protection.

The grinding process

Grinding is a metal-cutting process in which thousands of tiny cutting edges made of abrasive material are bonded together in a wheel, and the wheel is rotated at high speed and then brought into contact with the workpiece. The workpiece may be rotating or moving backwards and forwards (reciprocating) under the wheel. Grinding produces flat, cylindrical or shaped (formed) surfaces with a smooth surface finish and with high dimensional accuracy. It is particularly useful for machining hardened parts.

The bench grinder

This type of machine may be called a bench grinder, pedestal grinder or double-ended off-hand grinder. It is a general-purpose machine for fast metal removal when grinding small handtools or single-point tools. The workpiece is held in the hand and guided manually against the wheel (off-hand).

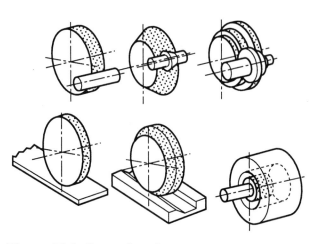

Figure 10.1 Ground surfaces.

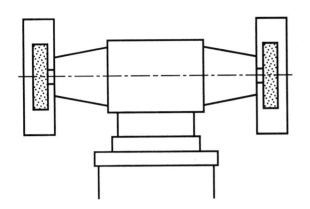

Figure 10.2 The bench grinder.

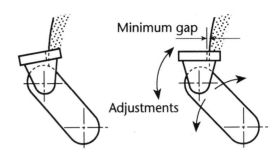

Figure 10.4 The adjustable tool rest.

This machine is basically an electric motor with a grinding wheel mounted either end of its central spindle. For convenience one wheel is coarse for very rapid metal removal, one wheel is fine to obtain a good surface finish.

Each wheel should have a protective screen for eye protection and a 'rest' to support the workpiece or the hand during grinding. These rests should be made of steel and securely attached to the grinder. They should be inspected regularly and adjusted so as to give the minimum clearance from the wheel. This prevents fingers becoming trapped or the workpiece getting jammed between the rest and the wheel, causing the wheel to break.

When using this type of machine do not force the workpiece against the wheel with very heavy pressure, do not grind on the side of the wheel, do not allow both wheels to be used at the same time and do not attempt to stop the wheels by applying pressure after the power has been switched off.

Wheels should be visually inspected for damage each day, and should be 'dressed' reg-

ularly to give a clean, flat surface for grinding, using a 'star' dresser.

Figure 10.5 Star dresser.

Wheels should be 'balanced' before they are mounted on the machine. Both dressing and wheel mounting must only be done by a competent person who has received specialist training.

The risk of a wheel bursting is highest when it starts up and it is good practice to stand to one side when first starting the wheel.

Always use eye protection when grinding.

The surface grinder

This machine is mainly used to produce flat surfaces using the periphery (edge) of a disc wheel or the end face of a cup-type wheel. The workpiece is securely clamped and passed beneath the grinding wheel.

There are various combinations of wheel type and workpiece motion for varying requirements.

Figure 10.3 Protective screen.

Horizontal spindle, disc wheel

Workpiece(s) reciprocating. Used mainly for long, flat surfaces, but can also produce angled faces and shoulders. It is rather slow. Generally used for precision work with small amounts of metal removal.

Workpiece(s) rotating. Used mainly for circular parts or thin rings (such as piston rings). Production is rapid. Machining lay or pattern is multidirectional which is good for mating surfaces.

Vertical spindle, cup wheel

Workpiece(s) reciprocating. There is a larger area of contact than with the disc wheel, giving higher rates of metal removal. Produces flat surfaces quickly and accurately.

Workpiece(s) rotating. Allows high rates of metal removal. Rotation of both the wheel and work generates a very flat surface. This process

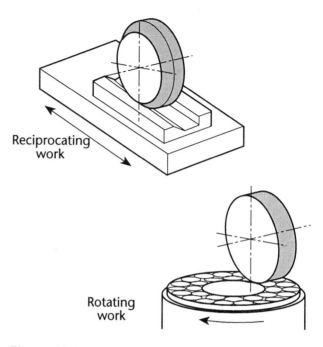

Figure 10.7 Horizontal spindle operations.

is good for rough uneven surfaces such as castings and forgings.

Cup-type wheels may be in the form of a cup, a ring or as a series of segments clamped to a circular chuck.

Segmented wheels are used where large amounts of metal are to be removed. It is cheaper to provide and replace large segments than a large solid wheel, and spacings provide for freer cutting.

Figure 10.6 The surface grinder.

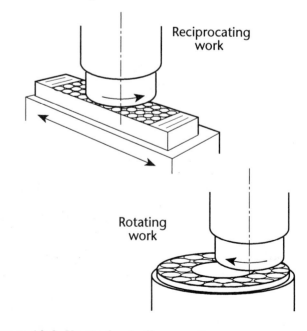

Figure 10.8 Vertical spindle operations.

Workholding

The most widely used workholding method on the surface grinder is the magnetic chuck.

Magnetic chucks

A magnetic field is created and used to hold ferrous workpieces tightly to the surface of the chuck. The workpiece is held by the frictional grip between the chuck and the workpiece surface.

The amount of holding force depends on the area of contact between the work and chuck, and the size and thickness of the workpiece. A smooth surface gives better holding power than a rough surface because there is more surface area in contact.

Very thin parts are not held securely because there are relatively few lines of magnetic force holding the part. To improve this condition, laminated blocks of brass and steel create more lines of flux passing through the part. This holds the part more securely even though the individual flux lines are weaker.

Electromagnetic chucks. These use a direct current (DC) flow of electricity passing around coils to create the field. The field may be varied in strength to suit the workpiece, the field may be 'shaped' by formed coils for special applications and residual magnetism can be removed from the workpiece.

A major disadvantage is the potential failure of the chuck if the power supply is interrupted.

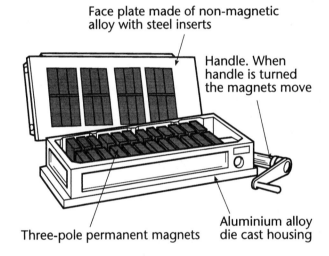

Face plate made of non-magnetic alloy with steel inserts

Handle. When handle is turned the magnets move

Three-pole permanent magnets

Aluminium alloy die cast housing

Figure 10.11 A permanent magnetic chuck.

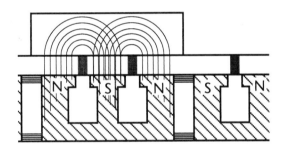

Figure 10.9 Magnetic fields.

Figure 10.10 Eclipse electromagnetic chuck.

Permanent magnetic chucks. These use three-pole permanent magnets to hold the workpiece. When not in use, 'keepers' or steel inserts in the chuck face contain the magnetic flux within the chuck. When in use, the magnets are moved away from the 'keepers' so that the flux extends above the chuck face and into the workpiece. An external handle is used to move the magnets.

The permanent magnet chuck does not need an external power supply, so it is safer. However, the magnetic force weakens in time, and also parts may need to be de-magnetised to remove residual magnetic charges.

Specially shaped chuck blocks are used to hold parts which cannot lie flat on a normal chuck, e.g. round parts or vee shapes.

Magnetic workholding is only suitable for ferrous parts (containing iron) and may leave the part with an induced magnetic field. De-magnetisers are used to remove these fields.

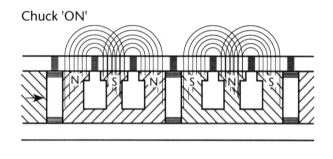

Figure 10.12 Permanent magnetic chuck.

Other workholding methods

Workpieces may be clamped directly to the machine table, held in machine vices or clamped to angle plates (to grind square faces).

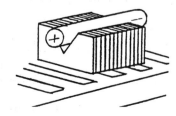

Figure 10.13 Chuck block.

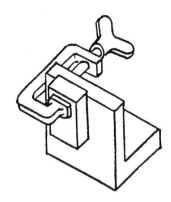

Figure 10.14 Angle plate.

Special fixtures may be designed for complex or high volume production.

Cylindrical grinders

The main types of cylindrical grinding machines are the plain grinder and the universal grinder.

The plain grinder

This machine is used for the production of external diameters, shoulders and tapers. Internal diameter grinding requires a special attachment.

The plain machine is very robust for the heavier cutting required in production grinding work. The workhead is fixed on the table and the wheelhead moves square to the table axis only.

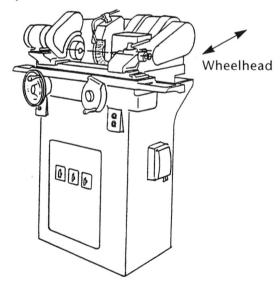

Figure 10.15 Plain cylindrical grinder.

The universal grinder

This machine is used for the more intricate and complex work required in the toolroom. The workhead may be swivelled on the table and the wheelhead may also be swivelled on its slide. The more complex machines have a compound slide which allows the wheel to be moved at an angle. These features allow large tapers to be ground on the workpiece.

The front face of the workpiece may be ground very flat using the periphery of the wheel, rather than the side of the wheel, in a generating process (see Fig. 10.17).

The universal machine has the facility to carry a second spindle for internal grinding operations.

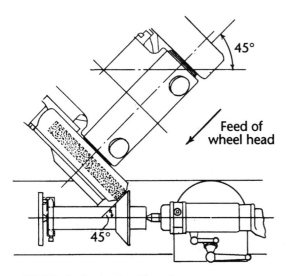

Figure 10.16 Swivel wheelhead.

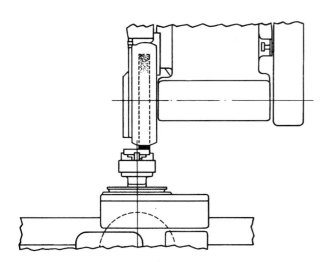

Figure 10.17 Swivel workhead.

Cylindrical grinding

External grinding

The workpiece is usually mounted on centres to ensure the production of concentric diameters. The workhead carries a centre which may be set as dead or live. The drive to the workpiece is through a carrier and catch plate arrangement. The carrier is usually of a different shape to that used on a lathe, to distribute the weight more evenly and reduce vibration. It is also lighter in weight. The driving dog is usually captive in the carrier and is adjusted to 'pull' the carrier rather than 'push' it, so reducing the amount of lost motion in the set-up.

The other dead centre is held in the tailstock and is springloaded to provide an adjustable clamping force and to allow for heat expansion of the part. Both centres are mounted on

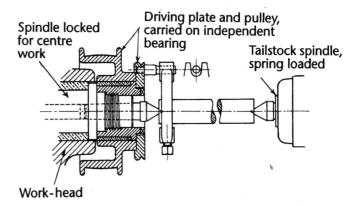

Figure 10.18 Grinding on centres.

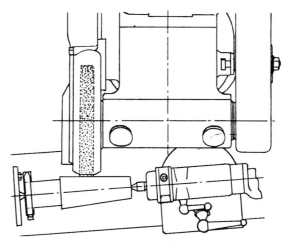

Figure 10.19 Grinding a taper using the swivel table.

a table, and this is reciprocated past the wheel so that long diameters may be generated on the workpiece.

The workpiece revolves against the rotation of the disc wheel. The table and workpiece can be swivelled so that accurate tapers may be generated. Very fine adjustment of the longitudinal movement of the table allows the side of the wheel to be used to generate flat shoulders.

On some machines, the wheelhead can be swivelled allowing coarse tapers to be plunge-cut on the workpiece. The wheel width is a limiting factor on the length of taper.

The wheel may be dressed for roughing (0.05 mm deep and fast traverse) or for finishing (0.012 mm deep and slow traverse). The wheel is set into the workpiece by about 0.05 mm at each pass when rough grinding and by 0.02 mm at each pass when finish grinding. When approaching final size this may be reduced to 0.01 mm per pass.

When using the side of the wheel to grind shoulders, minimal end feed must be used.

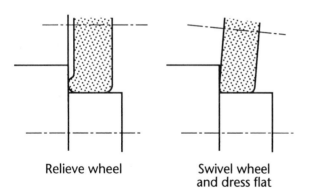

Relieve wheel   Swivel wheel and dress flat

Figure 10.20 Grinding shoulders.

Internal grinding

Particular care must be taken with internal grinding operations due to the large arc of contact, the high speeds and the relatively small wheel. Rapid wheel wear usually occurs and this must be taken into account when nearing finished work size. Plenty of coolant must be used.

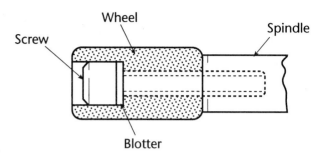

Figure 10.21 Internal wheel mounting.

Preparation

It is essential to ensure that all locating and mating faces are clean and that the spindle is running true, a dial test indicator may be used to check this. The correct type and size of wheel must be selected and securely mounted using a caphead screw and correct size blotter. Do not overtighten the screw.

The wheel is dressed for roughing (0.05 mm deep and fast traverse) or for finish grinding (0.012 mm deep and slow traverse).

If required, a dressing stick is used to dress the end face of the wheel.

Internal grinding operations

There are three main types of internal grinding operation.

Grinding a through bore. The wheel is set to overrun each end of the bore by one-third of the wheel length.

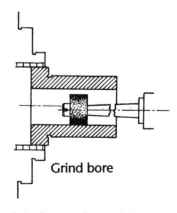

Grind bore

Figure 10.22 Grinding a through bore.

The wheel is first dressed for roughing and the bore ground to leave 0.05 mm on diameter. Ample coolant must be used and the bore must be checked for dimensional accuracy and parallelism. It may be necessary to adjust the workhead or table.

The wheel is then dressed for finishing and ground to finish size using small cuts of about 0.005 mm. Ample coolant must be used and the operator must wait for 'spark out'.

Special attention is required near the end of the operation due to wheel wear.

Grinding to a depth. This requires a recessed-type wheel with the screw head set well below the end of the wheel. The correct size of wheel must be used so that it is larger than the shoulder but must not grind on both sides of the centre hole.

The machine stops are set using a rule, allowing traverse to within 3 mm of the bottom of the bore and allowing one-third of the wheel to leave the bore.

The bore is rough and finish ground to size. The wheel is then dressed on its end face to give a small relief angle and 3 mm of land, and set 0.1 mm away from the finished bore.

With extreme care, the wheel is brought to touch the end face of the bore. Minimal end feed is used until the face has cleaned up and 'spark out' has occurred.

The wheel is redressed, set to within 0.01 mm of the finished bore and the end face ground as described above.

The final stages may require repetition and frequent redressing of the wheel to achieve the required dimensions.

Grinding an internal taper. Swivel either

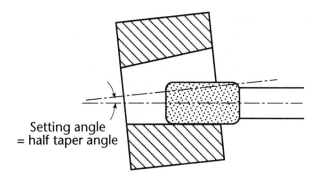

Figure 10.24 Grinding an internal taper.

the workhead or the table to half the included angle of the taper and then grind as a through bore. Check with a taper gauge.

Grinding wheels

The disc wheel normally cuts on its periphery. It is dressed flat and concentric to the spindle for most operations.

When the side of the wheel is to be used to grind shoulders, the area of contact must be reduced to avoid burning of the workpiece and undue sideways force on the wheel. The side of the wheel may be relieved with a dressing diamond, or a dressing stick, to leave a narrow band of about 3 mm. Alternatively, the head may be swivelled and the wheel periphery dressed flat so that the side of the wheel is at an angle to the face being ground (see Fig. 10.20).

The spindle

This must run true and without vibration. On

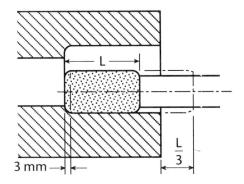

Figure 10.23 Grinding to a depth.

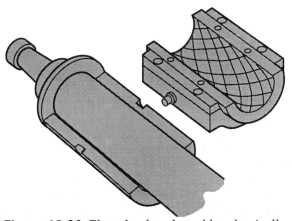

Figure 10.25 The wheel and workhead spindle.

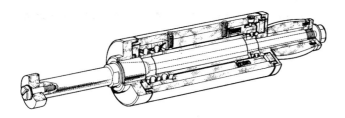

Figure 10.26 Internal spindle.

plain machines, the nitralloy spindle runs in white metal bearings, scraped by hand and pre-adjusted to remove any backlash.

On universal machines, air bearings may be used to reduce friction and heat generation or hard film microsphere bearings may be used with pressurised lubrication.

Internal grinding spindles are generally run on super-precision bearings which can carry both journal and thrust loads. These bearings are more suitable for the higher rpm required of internal operations.

Workholding

The most frequently used method of workholding is between centres. It is particularly important to ensure that centres in the ends of the workpiece are dead true and of the correct form. Errors may cause chatter, vibration and incorrectly ground forms. Hardened workpieces may distort due to quenching, in such cases the centres will have to be lapped to true size. Check centres for true form.

To set the work for parallel grinding, a parallel test mandrel is set on the centres (lubricated) and checked for runout over its length using a dial test indicator. The swivel table is adjusted to bring the mandrel parallel if required. The workpiece is then set up on the machine.

Attachment of the driving dog means that the whole workpiece cannot be ground without reversing the job.

In addition to the dead centre method of

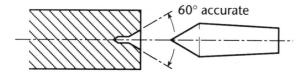

Figure 10.27 Workpiece centres.

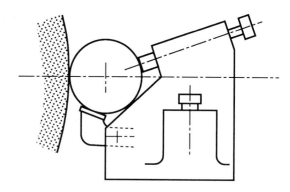

Figure 10.28 Grinding steady.

workholding, both three- and four-jaw chucks may be used, as well as magnetic chucks, collets and mandrels. When using mechanical chucks, allow the minimum amount of work to protrude from the jaws (minimum overhang) and ensure that the grinding wheel does not foul the chuck.

Too large an overhang gives rise to chatter marks and inaccurate dimensions. Overhang should be limited to twice (2×) the diameter of the work wherever possible.

Work is set parallel in the chuck using a dial test indicator and magnetic base, and lightly tapping the work with a mallet to achieve true running.

Grinding long, thin shafts may require the use of a two point steady to support the workpiece and reduce vibration and chatter. For production work, special fixtures may be used.

Safety

In addition to the general requirements for safety with grinding wheels, there must be adequate provision to guard the operator against excessive noise, vapour inhalation and splashing with coolant.

The internal grinding wheel is guarded by the workpiece itself during grinding but additional arrangements may be made on fixed internal grinding machines to guard the wheel when it is withdrawn from the work. Telescopic guards may be used.

The operator must take special care when internal grinding. Loading, unloading and checking all take place with the wheel adjacent to the operators body and great caution must be exercised.

11
Wheel selection

The grinding wheel has thousands of tiny cutting edges, in the form of abrasive grains, held together in a bonding material which also contains air gaps or voids. The cutting edges are discarded when blunt, giving a self-sharpening action to the wheel.

The most commonly used abrasives are aluminium oxide and silicon carbide, with cubic boron nitride, diamond and ceramic aluminium oxide also available.

Wheels are specified in terms of the type and size of abrasive, the type and strength (grade) of bond and the wheel structure. This information is given as a coded designation on the wheel. Factors affecting the final selection of the wheel include:

- the specific grinding process
- the workpiece material and its properties
- the required rate of metal removal and the specified surface finish
- the arc or area of contact between the workpiece and wheel
- the condition of the grinding machine

The wheel **shape** is governed by the type of machine, the mounting arrangement and the grinding operation.

The calculated speeds and feed of the workpiece and the wheel are adjusted to balance wheel life against the required surface finish and metal removal rate.

The grinding wheel

The grinding wheel is a special type of cutting tool composed of thousands of tiny cutting points. These points are formed by the sharp edges of abrasive grains (or grit) held in the wheel.

Each cutting point removes a tiny chip from the workpiece, just like a turning or milling tool. Eventually the cutting edge becomes blunt and it must bear a larger force in order to remove the chip from the workpiece. The force rises until it causes the grain to fracture and present a new, sharp edge to the workpiece. In this way the grain reduces in size until finally the cutting force causes it to be completely torn from the wheel, exposing new grains to the workpiece. These processes should ensure that the wheel is 'self-sharpening'.

Figure 11.1 The grinding wheel.

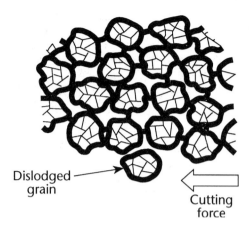

Figure 11.2 Self-sharpening action.

The main types of abrasive used in grinding wheels are aluminium oxide (75% of wheels used), silicon carbide and cubic boron nitride (CBN). Each abrasive has different properties of hardness and toughness which affect the manner in which dulling and fracture take place, in turn this determines which type of workpiece material is most suitable for a particular abrasive type.

The abrasive grains are held together by a bonding material. This material may hold the grains against high cutting forces (a hard bond) or may give up the grain under quite low cutting forces (a soft bond). When wheels are described as 'hard' or 'soft', it is this bonding strength which is being discussed.

Within the bonding material, there will be air gaps or voids. These provide spaces in which the tiny chips can lie and they help cooling fluids to reach the site of the metal-cutting operation. They also help to remove the heat from the cutting process.

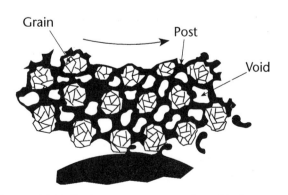

Figure 11.3 Basic components of a wheel.

Wheel specification

Type of abrasive

Aluminium oxide. A very hard material used to grind steels. The brown variety is used for high-strength materials such as alloy steels, high-speed steel and tough bronzes. The white variety fractures more easily giving a better self-sharpening action so it is used on hardened materials and for general toolroom grinding operations.

Aluminium oxide is not normally used on cast iron.

Silicon carbide. This is harder and tougher than aluminium oxide, it dulls and glazes rapidly with steels but gives a long wheel life with softer workpieces. The black variety is used for cast iron, bronze, aluminium, copper and non-metals.

The green variety is used to grind cemented carbides – suitable extraction equipment may be required.

Cubic boron nitride (CBN/Borazon). This is a manufactured abrasive, with a hardness between silicon carbide and diamond.

It is twice as hard as aluminium oxide and is cool cutting. It will withstand temperatures up to 1400°C before breaking down. It is used to grind very hard parts such as hardened steel rolls for rolling mills.

CBN wheels maintain their form over long production runs, require little dressing and do not burn the workpiece easily. These advantages compensate for the higher costs involved.

Diamond-bonded wheels. The wheel comprises a layer of diamond grains held in a non-abrasive core. They are used for grinding very hard or abrasive materials.

Applications include the sharpening of tungsten carbide cutting tools, grinding glass, ceramics, asbestos and items made of cement, and the cutting of silicon into wafers.

They need machines without play in the slides and with true-running spindles. They run at 1700–2000 m/min. The abrasive type designation will have the letter 'D' in the code.

Ceramic aluminium oxide. These wheels perform between aluminium oxide and CBN wheels. The ceramic content gives higher metal removal rates with longer wheel life and fewer dressing cycles required.

Norton Abrasives market these wheels as SG or Seeded Gel wheels. This denotes a new manufacturing process which gives a more uniform grain and a more controlled resharpening action.

Grain size

The grain size is expressed as a number. This number refers to the number of openings per linear inch in a mesh screen through which the grain is just able to pass.

For example: a 10 grain size abrasive will just pass through a screen with 10 openings per inch. The actual grain size in therefore 1/10-inch across. Grain sizes are grouped together under general classes as shown below:

- coarse 8, 10, 12, 14, 16, 20, 24
- medium 30, 36, 40, 46, 54, 60
- fine 70, 80, 90, 100, 120, 150, 180
- very fine 220, 240, 280, 320, 400, 500, 600.

Generally speaking, large coarse grains can remove more metal but give a rough finish. They are often used for hand or roughing grinders, and are best with soft, ductile materials like low carbon (mild) steel. If a large area of metal will be in contact with the wheel, coarse grains help to dissipate the heat that will be generated.

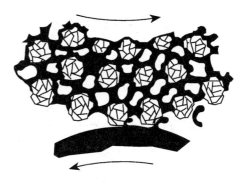

Figure 11.5a Large grains.

Fine grains remove smaller amounts of metal but give a better surface finish because there are more cutting edges across the wheels surface. They are best used on harder, brittle materials where the larger number of cutting edges are more effective.

If there is to be a small area of wheel and workpiece contact, with reduced heat generation, the larger number of cutting edges can be used for a better surface finish.

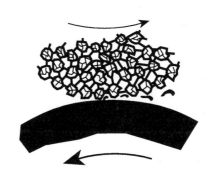

Figure 11.5b Small grains.

The grade of wheel

This is expressed as a letter and shows the degree of retaining 'grip' exerted on each grain by the bonding material and, therefore, the degree of cutting force needed to dislodge the grain.

Wheel grades are grouped together under general classes as shown below:

- very soft: E, F, G
- soft: H, I, J, K
- medium: L, M, N, O
- hard: P, Q, R, S
- very hard: T, U, V, W, X.

A soft grade of bond has a weak hold on the

Coarse grains

Medium grains
Actual grain sizes

ℬ 10

• 25

· 60

Figure 11.4 Sizing of grains.

abrasive grain. Blunt grains will be torn away easily, so the self-sharpening action will be pronounced. This is desirable when cutting hard metals as grains become blunt very quickly. Hence the term 'hard work – soft wheel'.

A soft grade is best with a large area of work and wheel contact which leads to rapid wheel wear. A soft grade can be used with large, rigid machines which do not impose additional forces on the wheel.

A hard grade of bond has a strong hold on the abrasive grain. It is used when cutting softer or weaker materials giving a longer wheel life. Hence the term 'soft work – hard wheel'.

A hard grade is best with a small area of work and wheel contact which causes less wear on the wheel. If the machine is not in very good condition, a harder grade may be required to withstand the additional forces imposed on the wheel by the machine operating characteristics.

If the grade is too soft, excessive wheel wear may take place. If the grade is too hard, the blunt grains may be retained for too long leading to a condition called **glazing** of the wheel.

A wheel may be made to act harder or softer by varying the forces acting on the grains. Decreasing the wheels speed or raising the feed rate, will increase the cutting forces. The wheel will shed grains, and wear quicker so it will appear to be acting as a softer grade of wheel. Increasing the wheel's speed (within its safe working speed limit) or slowing the feed rate, will decrease the cutting forces. The wheel will retain grains for a longer time period and appear to be acting as a harder grade of wheel.

The structure of the wheel

The structure is a measure of the relationship between the grain size and spacing, the bonding material and the voids (or spaces) in the wheel. Wheels can be manufactured to give specific structures ranging from very dense to very open. Structure is expressed as a numerical value between 0 and 15. General classes of structure are shown below:

• dense: 1, 2, 3, 4, 5
• medium: 6, 7, 8, 9
• open: 10, 11, 12, 13, 14, 15.

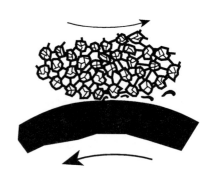

Figure 11.6 Dense structure.

The most important feature are the voids or spaces deliberately built into the wheel (sometimes referred to as the wheel porosity). These are designed to take the chips away from the machined surface, to avoid clogging or 'loading' the wheel face, and to allow grains to cut efficiently.

A dense structure has closely spaced grains and small voids. Only small amounts of material can be removed. It is used to achieve fine finishes on hard materials, to cut detailed or accurate forms and with small contact areas which produce limited amounts of chips.

An open structure can cope with high rates of metal removal so it is used for roughing operations on soft metals, and when a large contact area will give a large amount of chips. Carbides are normally ground with an open structure.

The bonding material

There are five bonding materials used to grip the abrasive grains. Each grain is held by a 'post' of bond material, in a similar way to a lathe tool in a toolpost. The post eventually fractures to release the blunt grain.

The type of bond material is related to the particular cutting operations and conditions that the wheel must cope with.

Vitrified bond. Used for 75% of grinding wheels. Clay is fused at high temperatures to give a hard, strong, glassy material unaffected by water, coolant or oils. High porosity and strength make it suitable for large rates of metal removal and accurate form work. These wheels are sensitive to impact.

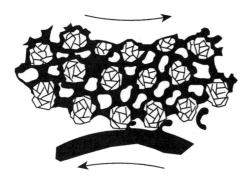

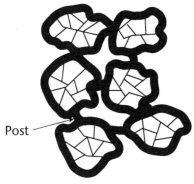

Post

Figure 11.7 Open structure.

Silicate bond. Silicate of Soda and clay are bonded together at a low temperature to give a lower strength bonding material.

Silicate bond wheels release the grains at lower cutting forces and are used to grind steel cutting edge tools, where a smoother cutting action is required and heat must be kept to a minimum.

Resinoid bond. Made from synthetic, thermosetting filled resins, to make 'elastic' wheels. They can be run at high speeds, cut cool and can accommodate some impact loading. They are used for cutting-off and for fettling castings (cleaning up).

Rubber bond. Made from vulcanized rubber (synthetic or natural) to give a strong, tough bonding material. It is ideal for very thin cutting off wheels and control wheels for centreless grinding.

Shellac bond. A naturally occurring resin, giving a bond of high strength with some flexibility. Porosity is low, so these wheels are used to give fine finishes on hard or tough steels. They may be used for slitting operations or work on thin sections where low heat generation is required.

Marking abrasive wheels – BS4481 part 1

Each wheel must be marked in a set way, to indicate the abrasive used, grain size, grade and bond. The maximum speed permitted for the wheel must be marked on all wheels larger than 55 mm diameter, smaller wheels must have the maximum speed displayed prominently at the machine. Additionally, manufacturers may include their own coding symbols.

Example of wheel marking

51 A 36 – L 5 V 23

51 the manufacturer's code for the abrasive being used

A   the abrasive code

    A   aluminium oxide

    S   silicon carbide

30 grain size – between 8 and 600

L   grade of bond – letters A to Z

5   structure – between 1 and 15

V   Bond material code
    V   vitrified
    S   silicate
    B   resinoid
    R   rubber
    E   shellac or other

23 manufacturer's code to record manufacture details.

Some manufacturers omit the structure letter from the wheel marking.

Grinding wheel shapes

Wheels for peripheral grinding

These wheels are used primarily on surface, cylindrical or off-hand grinding machines. Large sideways forces can cause wheel breakage and should be avoided, but light side cutting such as shoulder or form grinding is possible.

Wheels in this category include:

- straight or plain wheels
- recessed wheels
- offset wheels
- dovetail wheels
- tapered wheels.

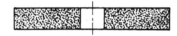

Figure 11.8 Plain wheel – type 1.

Recessed and offset wheels allow for clearance over the mounting flange and nut, and for mounting wheels that are wider than the mounting width.

Tapered wheels are used with portable grinders where the whole wheel cannot be guarded, and the extra strength given by the increasing cross section reduces the risk of bursting.

Cutting-off wheels are a special type of peripheral grinding wheel. They are of thin section, generally resinoid or rubber bonded, and may be flat or have a depressed centre. Only reinforced wheels may be used on portable machines or where workpieces are fed into the wheel by hand. Non-reinforced wheels may only be used on fixed machines which are adequately guarded.

Wheels for side grinding

Side grinding wheels may be found on machines such as the vertical spindle surface grinder, tool and cutter grinders and saw sharpening machines.

They are used with portable grinders for snagging operations – removal of rough edges and projections from castings, weldments, forgings and rolled billets.

Wheels in this category include:

- ring or cylindrical wheels
- cup wheels
- dish wheels
- saucer wheels.

Dish and saucer wheels are used on tool and cutter grinding machines.

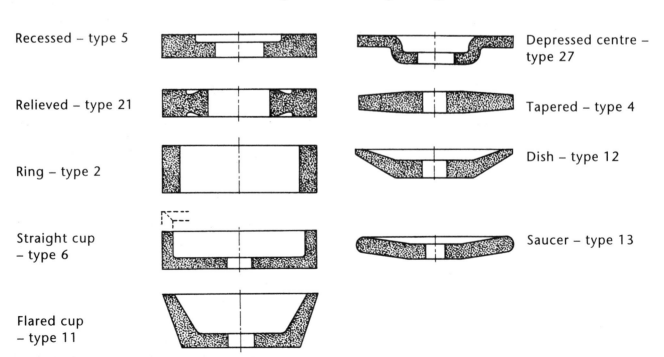

Recessed – type 5

Relieved – type 21

Ring – type 2

Straight cup – type 6

Flared cup – type 11

Depressed centre – type 27

Tapered – type 4

Dish – type 12

Saucer – type 13

Figure 11.9 Wheel shapes.

Segmental wheels

Segmental wheels are composed of many individual, replaceable segments. These may typically be 40 mm × 100 mm × 200 mm in size. They are mounted in a large holder so as to form an extensive grinding area for rapid metal removal. Generally used with vertical spindle machines.

Wheels for internal grinding

These may be in the form of non-mounted straight or recessed wheels, usually less than 100 mm diameter, and retained by a socket head screw, or as collet-held mounted wheels (a wheel formed around a small steel spindle).

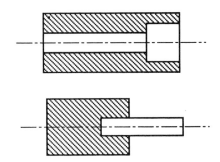

Figure 11.10 Internal grinding wheels.

Wheel selection

There are six main factors affecting the selection of the correct grinding wheel, these are:

- workpiece material
- workpiece properties and requirements
- metal removal rate
- surface finish
- arc and area of contact
- machine type and condition.

Workpiece material

Select aluminium oxide for high tensile strength materials such as:

- plain carbon steel
- alloy steels
- high-speed steel
- wrought iron
- malleable iron
- tungsten.

Select silicon carbide for materials with low tensile strength such as:

- cast irons
- brass and bronze
- aluminium
- tungsten carbide
- glass.

Workpiece properties

Hard materials – fine finish. Soft grade for rapid self-sharpening. Small grains for good finish and low metal removal. Dense structure to support fine grain.

Brittle materials. Small grains to provide many cutting edges.

Soft and ductile materials. Hard grade for long wheel life. Large grains for high metal removal but with poorer finish. Open structure to cope with metal removal rate.

Cemented carbides. Open structure for cool cutting.

Metal removal rate

High rate of metal removal – large grains, open structure, resinoid bond if appropriate.
    Low rate of metal removal – small grains and a dense structure.

Surface finish

Fine finish – small grains. Rough finish – large grains. Vitrified bond for good finish. Resinoid, rubber or shellac bond for fine finishes.

Arc and area of contact

The **arc** of contact is that part of the wheels diameter in contact with the workpiece. It varies with the following items:

- type of operation
- relative wheel/work sizes
- amount of wheel infeed.

The **area** of contact is formed by the arc of contact and the width of wheel in contact with the workpiece.

Arc of contact and type of operation. External cylindrical grinding has the smallest arc of contact. Surface grinding has a larger arc of contact. Internal cylindrical grinding has the largest arc of contact. Vertical spindle surface grinders using cup or ring wheels have variable **areas** of contact – there is no **arc** of contact.

Arc of contact and wheel size. A large diameter wheel has a longer arc of contact than a small diameter wheel.

Arc of contact and rate of infeed. The higher the rate of wheel infeed, the longer the arc of contact.

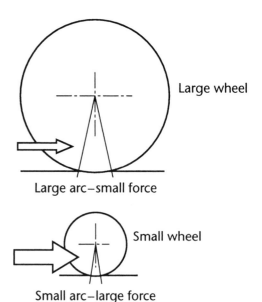

Figure 11.13 Arc of contact, wheel size and loading.

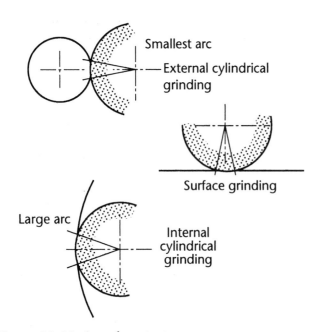

Figure 11.11 Arc of contact.

Arc of contact and wheel selection. In general, a large arc of contact provides more wheel length for the cutting operation, so the load on each grain is reduced. The wheel lasts longer and is said to act harder.

A small arc of contact means that fewer grains are available for cutting, so the load on each grain is increased. Wheel wear is increased and the wheel is said to act softer.

A large arc of contact produces more chips and heat and requires an open structure and larger grains.

A small arc of contact requires smaller grains and a dense structure.

Machine type and condition

Large, rigid machines maintained in good condition provide good support for the wheel during grinding. Machines which are poorly maintained, lack rigidity or have excessive wear, impose added stresses on the wheel and a harder grade may be required to give support to the grains. High power machines may require harder grades to cope with the more arduous cutting conditions.

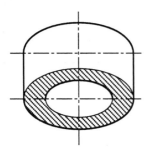

Figure 11.12 Area of contact.

Wheel and work speeds

The wheel manufacturer marks each wheel with the recommended maximum peripheral speed, this must not be exceeded.

In general, vitrified bond wheels work up to 35 m/s and resinoid, rubber or shellac bond wheels between 35 m/s and 80 m/s.

Work speed recommendations tend to vary between 30–120 m/min.

As the workpiece travels faster, each grain is required to remove more material. The wheel wears more rapidly and is said to act softer. More rapid metal removal is usually accompanied by a poorer surface finish.

Reducing the work speed reduces wheel wear, making the wheel act harder and improving surface finish.

Varying wheel and work speeds

Adjusting the wheel speed (within the manufacturer's stated limits) allows adjustment to the self-sharpening action of the wheel.

A faster wheel speed allows more grains to contact the workpiece, so reducing the load on each grain. Wheel wear is reduced and the wheel acts harder.

A slower wheel speed increases the cutting forces, wheel wear increases and the wheel acts softer.

Adjusting the work speed acts in the opposite way. A faster work speed forces each grain to cut more metal, wearing the wheel more quickly. The wheel acts softer.

A slower work speed reduces the metal cut by each grain and the wheel lasts longer. The wheel acts harder.

So:

- faster wheel speed – wheel acts harder
- slower wheel speed – wheel acts softer
- faster work speed – wheel acts softer
- slower work speed – wheel acts harder.

Grinding wheel speeds

These are generally of the order of 2000 m/min (6500 ft/min) for external grinding and 1500 m/min (5000 ft/min) for internal grinding.

The spindle speed is calculated from the wheel diameter using the formula:

$$N \text{ (rpm)} = \frac{\text{Surface speed (m/min)} \times 1000}{\pi \times \text{wheel diameter } D}$$

Note: The wheel must never be run in excess of the manufacturer's maximum recommended speed for that diameter.

Work speeds

These vary with the type of workpiece material, its condition and the grinding operation taking place. Some recommended values are shown below.

Internal grinding requires faster work speeds, up to 50% faster.

As a starting point use 25 m/min for external operations and 35 m/min for internal grinding.

The workhead spindle speed is calculated in a similar manner to the wheel spindle but using the workpiece diameter instead of the wheel diameter.

Feed rates

External grinding. For external grinding, the traverse feed rate is related to the wheel width. This is to reduce the effect of wear on the flatness of the wheel.

A traverse rate of two-thirds the wheel width per revolution causes the centre of the face to be worn concave because it cuts twice as much as each side.

A traverse rate of one-third the wheel width per revolution causes the edges to wear more than the centre giving a convex wheel.

A traverse rate of one-half the wheel width per revolution is the preferred rate to leave the wheel as flat as possible.

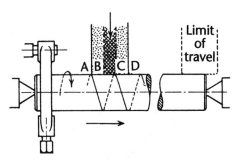

Figure 11.14a Traverse = 2/3-wheel width. Most wear at centre of wheel.

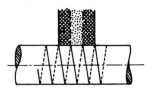

Figure 11.14b Traverse = 1/3-wheel width. Maximum wear concentrated at shaded portion of wheel face.

General recommendations for external grinding are:
- roughing soft steel – half of the wheel width per revolution of the work
- roughing hard steel – a quarter of the wheel width per revolution of the work
- finishing soft steel – a sixth of the wheel width per revolution of the work
- finishing hard steel – an eighth of the wheel width per revolution of the work

Internal grinding. For internal work, a high traverse rate is used to help spread the heat that is generated and to make the wheel act softer so as to help the selfsharpening process.

A traverse rate of up to one wheel width per revolution of the work may be possible. Specific cases vary with the type and condition of the work, the arc of contact and the rate of infeed.

Infeed of the wheel. This varies with the type of workpiece, the grinding operation, the rigidity of the machine and other factors. The infeed may vary between 0.005 and 0.04 mm.

When the wheel is fed into the workpiece, some distortion of the wheel or work takes place. Gradually, as grinding takes place, this distortion is removed. When no more sparks can be seen the distortion has gone, this is called 'spark out'.

If the grinding operation ceases before spark out occurs, the work will not be to the size indicated by the machine handwheels.

12

Wheel care

Wheels should be balanced before and after initial mounting and truing, and dressed to suit the particular grinding process. These operations should only be done by a trained and authorised person.

Wheels should be monitored for faults such as glazing and loading and corrective action taken. Guards of the correct type must be used and correctly adjusted to protect the operator and others, and to avoid damage to the wheel.

An appropriate type of coolant must be selected and used to aid the machining process and to remove dust and chips from the grinding area.

Wheels received from the manufacturer should be examined to detect any damage and checked to ensure that the wheel specification matches the original order.

Wheels must be handled and transported with great care to avoid accidental damage. A suitable storage area must be provided and wheels must be stored in a manner appropriate for their shape, size and type of bond.

Some types of wheel must be used within a certain time limit.

Wheel preparation

Wheel mounting (straight wheel)

1 First check that the wheel is sound, that bushings are in good condition and in place and that the wheel turns freely on the spindle without binding. Remove loose grains and any other similar material.
2 Assemble two flanges of equal diameter, recessed so that only the outer portion bears on the wheel. The flange diameter should be no less than one-third of the diameter of the wheel.
3 Also assemble two 'blotters'. These must be of compressible material, usually made of paper, between 0.3 and 0.8 mm thick. They should be slightly larger than the flanges, free from wrinkles and a good fit over the spindle. The purpose of the blotter is to dis-tribute the clamping pressure evenly over the wheel.
4 Place the inner flange on the spindle, over the key.

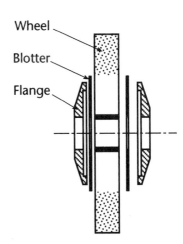

Wheel

Blotter

Flange

Figure 12.1

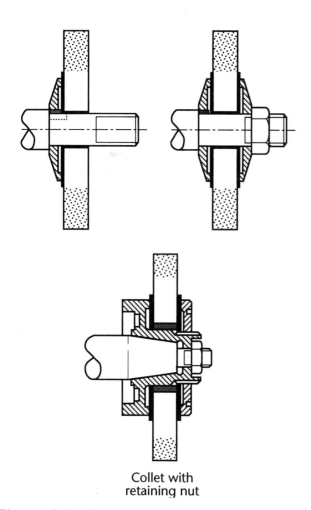

Collet with
retaining nut

Figure 12.2 Wheel assembly.

5 Mount the wheel on the spindle and push it snugly against the flange with a blotter between them. Check that the wheel and flange bear uniformly against each other.

6 Place the second blotter against the wheel and assemble the second flange to the wheel. Check again for uniform bearing, and that the flange slides easily over the spindle.

7 Tighten the spindle nut sufficient to hold the wheel without slippage, but not so tight as to set up excessive stresses in the wheel.

Some wheels have large holes and require special adaptor flanges for mounting.

Some grinding machines have a taper spindle. The wheel is mounted on a special collet which locates on the taper and is retained by a spindle nut (see Fig. 12.2).

Some collets comprise a hub flange which locates on the taper spindle, and a straight flange which locates on the hub and is secured

to it by a series of cap screws. Blotters are placed between the wheel and the flanges.

These collets often have an integral set of balance weights.

A set of wheels with different grades and grits can be assembled, each on its own collet. This makes wheel changing much quicker.

Wheel balancing

If the wheel is out-of-balance, it will cause vibration, chatter marks, damage to the spindle bearings and may even cause a wheel burst. The chatter marks produce a chess-board pattern and this is a good visual indicator of out-of-balance in the wheel. Both new wheels and wheels which have worn will need to be balanced.

Procedure.

1 The wheel is mounted on a balancing mandrel and this is supported on two knife edges, set perfectly horizontal. The wheel is free to rotate with minimum friction, and it will turn until the heaviest portion is at bottom dead centre (BDC). Mark this BDC point.

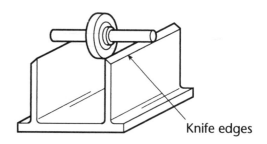

Knife edges

Figure 12.3 Wheel balancing.

2 Turn the wheel until the BDC mark is at the top point. Release the outer flange and allow the wheel to move downwards on the mandrel so that any clearance is at the bottom of the mandrel, opposite the heavy point. This will help balance the wheel before any weights are used.

3 The moveable balance weights are then set at 90° to the vertical line through BDC, and the wheel allowed to rotate again. A new low point will be created and its position should be marked.

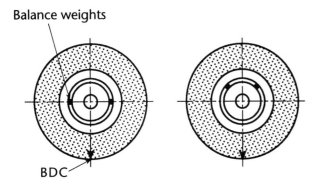

Figure 12.4 Adjusting the balance weights.

4 The balance weights are re-set equally about the vertical line through the new BDC, and they are gradually adjusted up or down together until the wheel will not rotate from any set position.

5 The wheel is now balanced and may be mounted on the machine spindle. It is a common practice to mount it with the marked heavy point at top-dead-centre so that any clearance between the spindle and bush brings the heavier side closer to the spindle centre. The machine guards are fitted securely before the machine is started up, and it is then run for about a minute to ensure that it is not about to burst.

Truing the wheel

After balancing, the outside face of the wheel and a portion of each side of the wheel is cleaned up with a diamond in order to bring these faces true to the running axis of the wheel.

With new wheels, rebalancing may be required after the initial truing operation.

When the wheel has worn away, the mass distribution may have changed and rebalancing and possibly retruing may be required.

During use, the wheel face may be worn unevenly and truing will be required to return it to a flat, square form.

Dressing the wheel

Dressing is an operation to change the nature of the cutting action of the wheel. It is required to remove dulled grains, to remove loaded grains and to alter the openness of the cutting face.

Figure 12.5 Star dresser.

Dressing may be done with a Huntington or star dresser on coarse grit wheels, used for off-hand grinding or snagging.

Precision grinding machines use a diamond, mounted in a suitable holder, for dressing operations.

Production machines may use a silicon carbide wheel to give rapid generation of a smooth, clean cutting face.

Dressing sticks are manipulated by hand for forming profiles and for dressing thin-section wheels.

Precautions with diamond dressers.

1 Cant the diamond at an angle to the axis of the wheel, and ensure that the contact point will not allow the diamond to 'dig in'. Cant angles are usually 10–15°.

2 Periodically, rotate the diamond in its holder to present a new facet. This keeps the diamond sharp. A blunt diamond will crush the grains, with some only partly fractured.

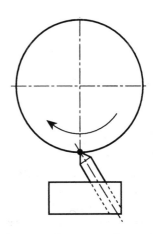

Figure 12.6 Diamond cant angle.

These will cause scratching of the work-piece. A dull diamond produces a dull wheel.

3 Make sure that the wheel does not grind into the diamond setting, unless the holder is of a type which is expressly designed to allow this, e.g. Multiset. The setting is usually a steel nib or a sintered matrix.

4 Always use plenty of coolant to prevent the diamond overheating and fracturing.

5 Make sure that the initial contact with the wheel is at its highest point, this may be at the wheel centre.

6 Use a low infeed after each pass. Usually 0.025 mm (0.001") but may be less with fine grain wheels. The rate of crossfeed must be related to the grain size, between 40% and 70% of the full grain size depending on whether a fine finish or an open face is needed.

Care-in-use of wheels

During use, wheels may be subject to loading or glazing. These conditions cause poor cutting conditions, generate excessive heat and can contribute to wheel breakdown or even bursting.

Loading

The cutting face becomes clogged with metal chippings and takes on a dirty appearance. The grains cannot cut freely. The wheel overheats and wears rapidly.

This condition is often associated with soft ductile materials or trying to cut too fast.

Correction factors.
• Check that the abrasive is correct for the type of workpiece material, Silicon Carbide is generally used with soft materials.
• Reduce the feed or speed of the work in order to reduce the metal removal rate.
• Reduce the wheel speed in order to make it act softer, replacing the cutting face before it can become loaded.

Glazing

Blunt grains have not been discarded but have been retained in the cutting face of the wheel. The wheel has a smooth and glazed appearance, and it rubs rather than cuts the work leaving burn marks. This is often a sign that the grade of wheel is too hard.

Correction factors.
• Replace with a softer grade of wheel.
• Increase the work speed to make the wheel act softer, discarding grains as they become blunt.
• Reduce the wheel speed to make the wheel act softer, discarding grains as they become blunt.

Figure 12.7 Loaded wheel.

Guarding grinding wheels

All abrasive wheels carry a risk of bursting when in service, and throwing wheel fragments into the area surrounding the wheel, causing serious injury to operators and bystanders.

An additional risk is that the operator may accidentally touch the rotating wheel and suffer serious injury. Such injuries may occur without the operator being really aware of what is happening, because often there is no pain at the time of the injury.

An unprotected wheel may be knocked and damaged when workpieces are set up or removed. Such damage may then cause the wheel to be weakened and to burst when in use.

For all these reasons it is necessary to provide suitable guards of adequate strength for grinding operations. Exceptions are mounted wheels and points and some wheels used for internal grinding.

Fitting a suitable guard may also avoid the fitting of a wheel which is too big for safe use.

Types of guard

Hood guards

Those are used with disc-type wheels which cut on the periphery. The guard should expose as little of the wheel as possible, and adjustable pieces may be incorporated to deal with wheel wear. Fastenings and anchorages must be strong enough to contain a bursting wheel.

Band guards

These are used with cup-type wheels and segmented wheels, which are end cutting. The guard should be no more than 25 mm bigger in diameter than the wheel, and should be adjustable in length to reduce wheel exposure to the minimum.

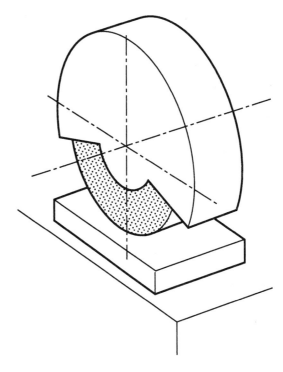

Figure 12.8 Hood guard.

Guard material

For most applications, guards will be made of steel castings or structural steel. For some low-energy applications (such as low speed, small bench grinders and thin section cutting-off wheels) cast iron may be used. Other materials may only be used after bursting trials have proved them adequate. Specifications for guard thickness, wheel speed and guard material are given in the relevant British Standard.

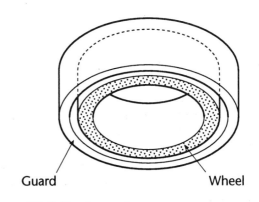

Guard   Wheel

Figure 12.9 Band guard.

Exposure angles for guards

The guard should expose as little of the wheel as possible. This exposure will vary with the kind of grinding operation being used, and examples of recommended maximum exposure angles for various operations are given below.

Guards should be capable of adjustment to minimise exposure (except for certain guards used on some portable machines).

Protective screens

Additional protection should be provided to guard the operator against coolant spray and workpieces which might be thrown out of the machine. Steel screens are usually placed around the work table for this purpose and they provide some extra protection in the event of a burst wheel.

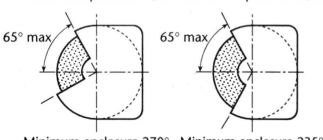

Bench and pedestal m/cs   Bench and pedestal m/cs

65° max   65° max

Minimum enclosure 270°   Minimum enclosure 235°

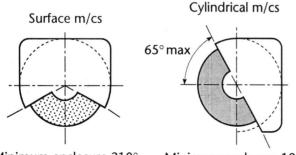

Surface m/cs   Cylindrical m/cs

65° max

Minimum enclosure 210°   Minimum enclosure 180°

Figure 12.10 Exposure angles.

Grinding fluids

Grinding fluids are primarily designed to cool and lubricate.

Types of grinding fluid

1 Chemical coolants are composed of water for cooling and chemical agents to reduce rusting and bacterial growth.
2 Soluble oils combine water for cooling and an oil emulsion for lubrication.
3 Straight mineral oils are mainly used for lubrication of high metal removal operations and where accuracy, surface finish and wheel life are important. These oils often need to be cooled.

Grinding fluid requirements

Cooling. This is required to reduce workpiece distortion and aid accuracy of form. Cooling reduces the tendency for chips to distort, so helping the cutting process, and cool chips reduce the problem of wheel loading.

Dust and chip removal. A good flow of coolant reduces the amount of airborne dust, which can be a health hazard. The level of dust in the atmosphere should not exceed 10 mg/m^3.

Coolant also conducts the chips away from the wheel and cutting area, reducing the problem of loading.

Application methods

The most common method of application is by flooding the contact area. A large volume of coolant is required, rather than a fast stream, to conduct heat away. 'Through the wheel' delivery places the coolant into the contact area through a dovetail groove in the wheel flange.

Coolant properties

The coolant must be stable, not easily contaminated and should not form a foam which would reduce its cooling qualities.

The coolant should not corrode or attack the elements of the grinding machine.

Precautions with fluids

Coolants can reduce the strength of resinoid, rubber and shellac wheels. The concentration and alkalinity should be checked regularly and the pH value of 8 should not be exceeded.

Coolant supply should be shut off and allowed to drain away after completing grinding operations and before stopping the machine. This ensures that wheels do not absorb coolant in one area causing out-of-balance conditions when restarted.

Receiving grinding wheels

Initial checking and unpacking

On receiving a wheel from the manufacturer, the container it comes in should be examined for signs of damage in transit.

The wheel should be carefully unpacked making sure that any tool used to open the container does not cause damage to the wheel.

The wheel should be cleaned with a brush and visually examined for cracks and chips.

The size, grit grade and bond should be checked against the despatch note and the original invoice.

Examination for soundness

The soundness of a vitrified wheel can be further checked by suspending it vertically and tapping it with a non-metallic instrument

Figure 12.11 Examination for soundness.

such as a wooden mallet. A clear ringing tone should be produced. A 'dead' sound indicates a cracked wheel and it should not be used.

The wheel should be tapped at two points 45° from the vertical and about 30–40 mm from the periphery. The wheel is then rotated by 45° and the test repeated.

Larger, thicker wheels may be tested on the periphery rather than on their sides.

Organic bonded and filled wheels do not produce a clear ringing tone, but the sound produced is still a good indicator of damage to the wheel.

Handling abrasive wheels

All grinding wheels are relatively fragile and must be handled with care.
1 Take care not to bump or drop wheels. Do not allow anything to fall onto a wheel.
2 Wheels which cannot be carried by hand should be moved by a truck or conveyor.
3 If wheels are too big to move in the ways described above, they may be rolled into position. A soft resilient surface must be provided all along the wheels path.
4 Wheels must be carefully stacked for security during transit. They must not be able to topple over, and heavy items must not be stacked on top of them.

Storing grinding wheels

The storage area. Wheels must be stored in a warm, dry place in suitably designed racks. They should be under the supervision of a competent individual.

Storage methods.

1 Small wheels (less than 80 mm diameter), mounted wheels and points may be stored in draws, bins or boxes. Exceptions include flared cup, dish and saucer shape wheels.
2 Wheels for peripheral grinding can be stood on edge with two-point support in wooden racks. They must not be able to tip over. Exceptions include dish and saucer type wheels.
3 Cylinder wheels and straight cup wheels of thick rim and hard grade may also be stored on their edge as above.

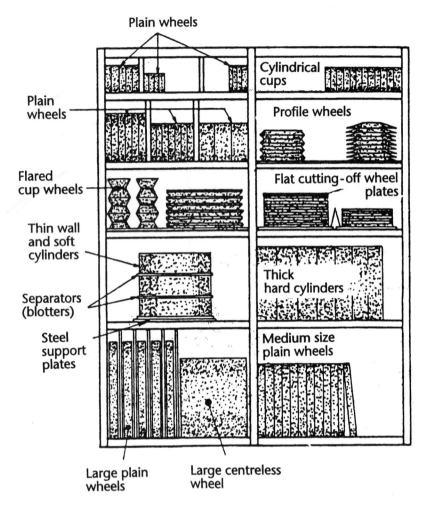

Figure 12.12 Storage of grinding wheels.

4 Cylinder wheels and straight cup wheels of normal thickness, together with dish and saucer wheels should be laid on their flat sides and stacked with some form of cushioning material between them.

5 Soft grades of straight cup wheels, and all flared cup wheels, should be stacked base-to-base and rim-to-rim to avoid chipping the edges and cracking the side walls. They should not be stacked one inside another.

6 Thin resinoid, rubber and shellac wheels should be stored flat on a rigid surface to prevent them from warping during storage.

Wheels with a limited lifetime. Resinoid, rubber, shellac and some bonded wheels deteriorate over time. They should be used within two years of receiving them.

13
Self-assessment questions

Chapter 1. Health and safety

1 Who is responsible for safety in workshops?
2 Give two examples of omissions which could lead to accidents.
3 Explain the difference between an Improvement Notice and a Prohibition Notice issued by a Health and Safety Inspector.
4 List four types of substance which could affect health and safety.
5 List three methods of avoiding or preventing direct contact with dangerous parts of a machine.
6 List four controls which should be present on all machines.
7 State two precautions to be taken when moving a load using a two-person lift.
8 List four requirements of workshop floors under health and safety legislation.
9 List three types of personal protective equipment which would help prevent eye injuries.
10 Describe two provisions of the Abrasive Wheels Regulations which are still law.
11 List six general safety precautions for machine operators.

Chapter 2. Metal cutting

1 Describe the condition of the chip and the workpiece when the cutting speed is too fast.
2 a) Sketch tools which have:
   i) positive rake
   ii) negative rake
   iii) zero rake.
   b) State which tool carries the largest cutting force.
3 Give two reasons for using negative rake toolholders with tungsten carbide inserted tips.
4 Explain why a boring tool requires a larger clearance angle than a standard turning tool. Use a sketch to illustrate your answer.
5 Calculate the spindle speed when turning aluminium alloy of 25 mm diameter at 90 m/min.
6 State two materials which produce continuous chips.
7 State two effects of a built up edge.
8 List three materials producing discontinuous chips.
9 List four reasons for using chipbreakers.
10 Sketch three types of ground-in chipbreaker. State one limitation of this type of chipbreaker.
11 List four reasons for using cutting fluids.
12 Explain the difference between:
   a) soluble oil
   b) straight mineral oil
   c) EP oils.
13 List four precautions to be taken when using or preparing cutting fluids.

14 Describe the reasons why cutting fluids are sometimes not used with carbide tools.

Chapter 3. The lathe

1 List six basic requirements of a lathe.
2 State two merits of using grey cast iron for a machine body.
3 List four safety factors related to the machine power supply.
4 State two different methods of transmitting motion which are used on the carriage.
5 Name the mechanism used to engage the leadscrew drive.
6 State the advantage of using a vee-flat type of guideway.
7 Describe one simple method of checking the alignment of the headstock and the tailstock.
8 Describe the main physical difference between the turret lathe and the capstan lathe.
9 Compare the types of workpiece which are most suitable for the turret lathe and the capstan lathe.
10 Explain why the CNC lathe has no handwheels or graduated dials.
11 List two advantages of using ballscrews rather than conventional leadscrews on CNC lathes.
12 State the reason for a slant bed on a lathe.
13 Explain why CNC lathes have totally enclosed guards.
14 Describe the type of workpiece which is most suitable for CNC turning.

Chapter 4. Cutting tools

1 Explain what is meant by 'hot hardness' in a high-speed steel tool.
2 What quality of HSS makes it suitable for interrupted cutting?
3 List two types of HSS cutting tool and compare their advantages and disadvantages.
4 State one advantage and one disadvantage of cemented carbides over HSS cutting tools.
5 Select a typical carbide grade and number for finish turning mild steel.

6 List three benefits of using indexable inserts.
7 Sketch a right-hand light turning and facing tool. State the circumstances in which you would use this type of tool.
8 State four precautions to be observed when setting up to use a parting-off tool.
9 State the tool which is used to improve the finish produced by a screw-cutting tool.
10 List two types of knurling tool.
11 Explain the difference in application between a 80° rhomboid insert and a 35° rhomboid insert.
12 Sketch a tool having a 15° (positive) approach angle.
13 When turning an external diameter, the tool is set below the workpiece centre
   a) state the effect on the turning operation
   b) state the effect this setting has on the working rake angle.
14 State two advantages and one disadvantage of oblique cutting when compared to orthogonal cutting.
15 Give two reasons why carbide tools are set to produce a short and fat chip rather than a long and thin one.
16 State three possible causes of flank wear on a carbide tool.

Chapter 5. Lathe workholding

1 List three methods of fastening universal chucks to the lathe.
2 Describe two types of workpiece which are suitable for holding in a self-centring chuck and two types which are unsuitable.
3 Describe two machining applications for which the independent four-jaw chuck would be suitable.
4 List three types of collet used on lathes.
5 State two advantages and two limitations of collet chucks.
6 Describe two types of machining operation which are often performed on a face plate.
7 Describe the process of balancing a face plate.
8 List two reasons for using a fixed steady.
9 Explain the reason for using a travelling steady when turning a long shaft.

10 State the reason for turning work on a mandrel. List four types of mandrel.
11 State the circumstances where a spigot would be used in preference to a mandrel.
12 Explain the purpose of bungs when turning on centres.
13 Describe the different applications of a live centre, a dead centre and a running centre.
14 State one problem which may arise when turning work on a running centre.

Chapter 6. Turning operations

1 Explain the difference between forming and generating operations.
2 Explain why a reaming operation is sometimes preceded by a boring operation.
3 Describe two methods of producing eccentric external diameters on the lathe.
4 List two faults associated with an incorrect centre drilled form and describe the effect of each fault.
5 State two reasons for reaming holes.
6 List four methods of turning tapers. State the advantages and disadvantages of each method.
7 Describe the process for checking a tapered part using a taper inspection gauge.
8 Compare vee threads and square threads in terms of ease of production, strength and frictional forces generated in their use.
9 State the class of fit given by a 7H/8g nut and threaded shaft combination.
10 Compare radial infeed and flank infeed methods of screw cutting in terms of tool loading, workpiece material and the type of thread produced.
11 Calculate the lead of a two-start thread of 1.5 mm pitch.
12 State the purpose of hand chasing a thread.
13 Describe the complete procedure for tapping a vee thread from the tailstock.
14 List the sequence for screw cutting a two-start square thread.
15 List six safety precautions to be observed when turning.

Chapter 7. Milling machines

1 List six typical machined features produced on a milling machine.
2 Describe one major feature which distinguishes the universal milling machine from the plain horizontal milling machine. Explain the use of this feature.
3 Describe one important alignment check for the vertical milling machine and one for the horizontal milling machine.
4 Sketch and label a double negative carbide face mill.
5 Compare the advantages and disadvantages of 45° and 90° approach angle carbide face mills.
6 Explain why turning and milling cutter inserts have differently shaped corners.
7 A ⌀16 mm slot drill is used to machine a tough steel at 25 m/min cutting speed. Calculate the spindle speed.
8 The cutter in question 9 is to have a feed per tooth of 0.15 mm. Calculate the feed/rev and the feed/min.
9 Make a sketch of a cutter machining with conventional or up-cut milling. Show the direction of rotation of the cutter and the direction of feed of the table.
10 Describe the type of workpiece materials for which up-cut milling is particularly suitable.
11 Explain why a 'backlash' eliminator is required when machining with down-cut milling.
12 List six safety precautions to be observed when machining on the milling machine.

Chapter 8. Milling workholding

1 Make a sketch of a strap clamp and show how the packing and stud should be positioned to give the best clamping arrangement.
2 List two methods of applying the clamping force other than by a stud or threaded fastener.
3 Explain why tenons or keys are sometimes fitted to the base of a vice.
4 List the sequence for milling four sides of a workpiece square to each other.

5 Describe the circumstances where a vee block may be used as a workholding device.

6 List three important features of a milling fixture.

7 State two types of machining operation which could be performed on a dividing head.

8 Explain the difference between direct indexing and simple indexing on the plain dividing head.

9 State the worm/wheel ratio normally found on a dividing head.

10 A plain dividing head has a collar with 24 serrations mounted behind the chuck. Describe how the workpiece may be rotated by 75°.

11 Calculate the indexing movement to index 45° using simple indexing.

12 A dividing head is supplied with a hole plate with hole circles of: 15, 16, 17, 18, 19, 21, 29, 33, 39, 43, 49, 50 and 54 holes. Calculate the index movement to provide 36 divisions.

13 State one machining operation which requires a universal dividing head.

14 State six safety precautions to be observed with milling workholding.

Chapter 9. Milling operations

1 List two features which identify arbor mounted cutters.

2 Describe two differences between a light duty and a heavy duty plain cylindrical cutter (slab mill).

3 Identify the features of a helical mill which distinguish it from a plain cylindrical cutter.

4 List three types of machining operation for which a helical mill would be suitable.

5 Describe the difference in appearance and use between a side and face cutter and a slitting saw.

6 Using sketches, describe how a form relieved cutter is reground.

7 List two precautions to be observed when removing and/or storing machine arbors.

8 Sketch a typical arrangement of milling cutters for the machining operation called gang milling. State one advantage and one disadvantage of such a setup.

9 Identify the additional item of equipment which may be required on the milling machine when gang milling.

10 Describe two differences between an end mill and a slot drill.

11 Identify the differences in tooth form between a woodruff cutter and a tee slot cutter.

12 Make a sketch of a collet chuck mounted in the machine spindle. Show how the chuck is located, and retained. State how the cutter is driven.

13 List the tools and procedure for milling a tenon slot in a workpiece.

14 List six safety precautions to be employed when milling.

Chapter 10. Grinding

1 State three important reasons for using grinding processes.

2 Explain why there are usually two wheels on the off-hand grinding machine.

3 Describe the particular hazards associated with incorrectly adjusted rests.

4 List four precautions to be observed when using the off-hand grinding machine.

5 Identify which persons may balance and mount grinding wheels.

6 Sketch the following surface grinding setups:
a) horizontal spindle/reciprocating work
b) vertical spindle/rotating work.
State the advantages and disadvantages of each setup.

7 List two types of magnetic chuck. State the advantages and disadvantages of each type.

8 State three factors which affect the holding force between a magnetic chuck and the workpiece.

9 Describe the major differences between a plain and a universal cylindrical grinding machine.

10 Describe two methods of grinding external tapers.

11 Describe two methods of grinding shoulders.

12 Explain the reasons why extra care must be taken when internally grinding accurate bores.

13 Explain the special safety requirements needed when internal grinding.

Chapter 11. Wheel selection

1 Describe the process of 'self-sharpening'.
2 List the three main types of abrasive. For each type, state the workpiece or material for which it is most suitable.
3 Explain the reasons for selecting:
   a) a small grain size
   b) a large grain size.
4 Explain what is meant by the 'grade' of a wheel.
5 Describe how a soft grade of wheel may be made to 'act' harder.
6 Explain why an open structure is used with a large grain size or a large area of contact.
7 Identify the type of bond material which is:
   a) the most widely used
   b) most suitable for shock loads.
8 A wheel is marked as follows: A 40 M 8 V. Explain what is meant by each letter and number.
9 Explain the reasons for using:
   a) offset grinding wheels
   b) tapered grinding wheels.
10 Sketch two types of wheel used for internal grinding.
11 List six factors affecting wheel selection.
12 State which of the grinding processes has the largest arc of contact.
13 State a typical peripheral speed for a vitrified bond grinding wheel.
14 Describe with sketches how the traverse rate affects the flatness of the wheel when external grinding.
15 Explain what is meant by 'spark out'.

Chapter 12. Wheel care

1 State the purpose of the blotters when mounting a wheel.
2 List four effects of an out-of-balance wheel.
3 Describe the typical pattern created on the workpiece by an out-of-balance wheel.
4 Explain the difference between truing and dressing the wheel.
5 List six precautions to be observed when dressing a wheel with a diamond dresser.
6 Explain the causes of 'glazing' and how the problem can be corrected.
7 State four reasons for guarding a grinding wheel.
8 State three types of material used for guards.
9 State the 'minimum enclosure' angle for:
   a) a cylindrical grinding machine guard
   b) a surface grinding machine guard.
10 List four properties that a grinding fluid should have.
11 Describe two ways in which a cutting fluid can adversely affect a grinding wheel.
12 List four precautions to be taken when checking and unpacking a wheel.
13 Describe the method of checking a vitrified wheel for soundness.
14 List three precautions to be observed when handling or transporting grinding wheels.
15 Describe how cylinder wheels should be stacked.

Index